1 MONTH OF
FREE
READING

at
www.ForgottenBooks.com

By purchasing this book you are eligible for one month membership to ForgottenBooks.com, giving you unlimited access to our entire collection of over 1,000,000 titles via our web site and mobile apps.

To claim your free month visit:
www.forgottenbooks.com/free911406

ISBN 978-0-265-92984-1
PIBN 10911406

NBS TECHNICAL NOTE 6

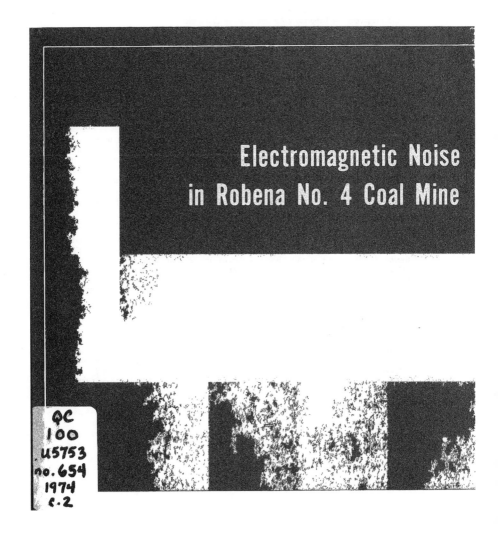

Electromagnetic Noise
in Robena No. 4 Coal Mine

NATIONAL BUREAU OF STANDARDS

The National Bureau of Standards[1] was established by an act of Congress March 3, 1901. The Bureau's overall goal is to strengthen and advance the Nation's science and technology and facilitate their effective application for public benefit To this end, the Bureau conducts research and provides (1) a basis for the Nation's physical measurement system, (2) scientific and technological services for industry and government, (3) a technical basis for equity in trade, and (4) technical services to promote public safety The Bureau consists of the Institute for Basic Standards, the Institute for Materials Research, the Institute for Applied Technology, the Institute for Computer Sciences and Technology, and the Office for Information Programs.

THE INSTITUTE FOR BASIC STANDARDS provides the central basis within the United States of a complete and consistent system of physical measurement; coordinates that system with measurement systems of other nations, and furnishes essential services leading to accurate and uniform physical measurements throughout the Nation's scientific community, industry, and commerce The Institute consists of a Center for Radiation Research, an Office of Measurement Services and the following divisions:

Applied Mathematics — Electricity — Mechanics — Heat — Optical Physics — Nuclear Sciences[2] — Applied Radiation[2] — Quantum Electronics[3] — Electromagnetics[3] — Time and Frequency[3] — Laboratory Astrophysics[3] — Cryogenics[3].

THE INSTITUTE FOR MATERIALS RESEARCH conducts materials research leading to improved methods of measurement, standards, and data on the properties of well-characterized materials needed by industry, commerce, educational institutions, and Government; provides advisory and research services to other Government agencies; and develops, produces, and distributes standard reference materials. The Institute consists of the Office of Standard Reference Materials and the following divisions:

Analytical Chemistry — Polymers — Metallurgy — Inorganic Materials — Reactor Radiation — Physical Chemistry.

THE INSTITUTE FOR APPLIED TECHNOLOGY provides technical services to promote the use of available technology and to facilitate technological innovation in industry and Government; cooperates with public and private organizations leading to the development of technological standards (including mandatory safety standards), codes and methods of test; and provides technical advice and services to Government agencies upon request. The Institute consists of a Center for Building Technology and the following divisions and offices:

Engineering and Product Standards — Weights and Measures — Invention and Innovation — Product Evaluation Technology — Electronic Technology — Technical Analysis — Measurement Engineering — Structures, Materials, and Life Safety[4] — Building Environment[4] — Technical Evaluation and Application[4] — Fire Technology.

THE INSTITUTE FOR COMPUTER SCIENCES AND TECHNOLOGY conducts research and provides technical services designed to aid Government agencies in improving cost effectiveness in the conduct of their programs through the selection, acquisition, and effective utilization of automatic data processing equipment; and serves as the principal focus within the executive branch for the development of Federal standards for automatic data processing equipment, techniques, and computer languages. The Institute consists of the following divisions:

Computer Services — Systems and Software — Computer Systems Engineering — Information Technology.

THE OFFICE FOR INFORMATION PROGRAMS promotes optimum dissemination and accessibility of scientific information generated within NBS and other agencies of the Federal Government; promotes the development of the National Standard Reference Data System and a system of information analysis centers dealing with the broader aspects of the National Measurement System, provides appropriate services to ensure that the NBS staff has optimum accessibility to the scientific information of the world. The Office consists of the following organizational units:

Office of Standard Reference Data — Office of Information Activities — Office of Technical Publications — Library — Office of International Relations.

[1] Headquarters and Laboratories at Gaithersburg, Maryland, unless otherwise noted; mailing address Washington, D C 20234
[2] Part of the Center for Radiation Research
[3] Located at Boulder, Colorado 80302.
[4] Part of the Center for Building Technology

Electromagnetic Noise in Robena No. 4 Coal Mine

W.D. Bensema

Motohisa Kanda

John W. Adams

Electromagnetics Division
Institute for Basic Standards
National Bureau of Standards
Boulder, Colorado 80302

'Technical note no. 654

Sponsored by
U.S. Bureau of Mines
Pittsburgh Mining and Safety Research Center
4800 Forbes Avenue
Pittsburgh, Pennsylvania 15213

National Bureau of Standards Technical Note 654

Nat. Bur. Stand. (U.S.), Tech. Note 654, 194 pages (April 1974)

CODEN: NBTNAE

For sale by the Superintendent of Documents, U.S. Government Printing Office, Washington, D.C. 20402
(Order by SD Catalogue No. C13.46:654). $1.50

CONTENTS

iii

CONTENTS (continued)

iv

CONTENTS (continued)

LIST OF TABLES

LIST OF FIGURES

LIST OF FIGURES (continued)

viii

LIST OF FIGURES (continued)

LIST OF FIGURES (continued)

LIST OF FIGURES (continued)

LIST OF FIGURES (continued)

LIST OF FIGURES (continued)

ABSTRACT

Two different techniques were used to make measurements of the absolute value of electromagnetic noise in
an operating coal mine, Robena No. 4, located near
Waynesburg, Pennsylvania. One technique measures noise
over the entire electromagnetic spectrum of interest for
brief time periods. With present instrumentation, the
spectrum can be covered from 40 Hz to 400 kHz. It is
recorded using broad-band analog magnetic tape, and the
noise data are later transformed to give spectral plots.
The other technique records noise envelopes at several
discrete frequencies for a sufficient amount of time to
provide amplitude probability distributions.
    The specific measured results are given in a number
of spectral plots and amplitude probability distribution
plots. The general results are that at frequencies below
10 kHz, power line noise within the mine is severe.
Impulsive noise is severe near arcing trolleys, and
at lower frequencies near any transmission line. Carrier
trolley phone signals and harmonics are strong throughout
the mine whenever the trolley phone is in operation.
    Additional information beyond that included in this
report may be obtained from the authors, who are with
the Electromagnetics Division of the National Bureau of
Standards.

Key words: Amplitude probability distribution; coal mine
noise; digital data; electromagnetic interference; electromagnetic noise; emergency communications; Fast Fourier
Transform; Gaussian distribution; impulsive noise; magnetic field strength; measurement instrumentation;
spectral density; time-dependent spectral density.

I. INTRODUCTION

The need for reliable communication systems in coal mines

is a long-standing problem. For emergency use, when all power

in a mine is off, the residual electromagnetic noise is no

problem. However, if a communication system were designed only

for emergency use, it would have two serious drawbacks. First, it would not be ready for immediate use in an emergency; second, it would not be of any value during normal operations. Therefore, the Bureau of Mines decided to design a communication system that could be used for both emergency and normal operational conditions.

During operation, the machinery used in a coal mine creates a wide range of many types of intense electromagnetic interference (EMI), and therefore ambient EMI is a major limiting factor in the design of a communication system.

The work reported here gives the results of the first comprehensive measurements of this EMI in critical communication locations such as working sections where miners extract coal.

There are several EMI parameters that can be measured: magnetic field strength, H; electric field strength, E; conducted current, i; and voltage, v, between two conductors. We made some measurements of each of these parameters, but one parameter was emphasized, magnetic field strength. There are several reasons. First, electric field sensors are notoriously insensitive at lower frequencies, and hence probably will not be useful in any practical mine communication system. Second, at any air-earth interface, only the magnetic field is essentially undisturbed, while the electric field is severely reduced. Third, any currents will induce magnetic fields, and hence measurement of the magnetic field will directly reflect currents. Fourth, trolley-wire noise voltages are propagated as transmission line phenomena, are directly related to transmission line currents, and hence to magnetic fields induced. Thus, measuring magnetic field strength gives a representative composite picture of noise from currents and voltages from most sources, as well as measuring the magnetic fields induced by arcing equipment.

As just mentioned, magnetic field strength measurements are emphasized, but even this one parameter is difficult to measure meaningfully. The IEEE definition [1] of magnetic field strength, H (magnitude of the magnetic field vector), is used in this report. Since there is a multitude of different sources that generate all known types of noise, the resultant magnetic field strength noise vector is a function of frequency, time, orientation, and location. Small variations in these quantities can cause many tens of decibels difference in measured field strength.

We used two measurement techniques. The first technique covers a large portion of the spectrum as a "snapshot" at one instant of time. In three-dimensional form, several such "snapshots" can show how drastically a signal varies over a period of a few seconds. The second technique records variations over a 20-minute time interval as a view at one frequency. We used a set of eight different frequencies. Both techniques measure three orthogonal components of magnetic field strength by using three systems simultaneously or by varying the orientation of one system; both techniques were used in as many different locations as possible within practical time limitations. Whether the noise signal tends to be Gaussian or impulsive depends on the number of sources and the distance to each source.

With the exception of the 3-D spectral plots, all measured noise is reported in absolute quantities (instead of relative) to allow others to make effective use of the data. For the magnetic field strength measurements, the NBS electromagnetic field calibration site was used with each complete measurement system to assure correct calibration.

A further complication in making these measurements is the need to have either permissible equipment or to use explosion-proof enclosures for non-permissible equipment. The mine environment is generally humid, dusty, poorly lighted, and without normal electrical power. We used battery-operated instruments for all of our portable measuring equipment.

There are two types of noise recorded in the spectral plots, and hence two different magnetic field strength parameters are required, H and H_d. Results are given as the rms value of one component of magnetic field strength, H, versus frequency for discrete frequencies, or magnetic-field-strength spectrum density level, H_d, [1] versus frequency for broadband noise in the spectral plots. Results are given as the rms value of one component of magnetic field strength versus percent of time this value is exceeded in the amplitude probability distributions (APD's). The rms value of an APD is representative of the actual peak value only as far as the measurement-system detector bandwidth will allow the detector to follow the time variations of the actual magnetic field. (In this context, noise envelope is sometimes used.) The results are applicable for a communication receiver whose bandwidth is similar to the measurement-system detector bandwidth.

Only representative samples of the total data measured are given in this report. Only a limited set of data-presentation formats have been used. If additional data, or data presentation in other formats, are required, please contact any of the authors at the Electromagnetics Division of the National Bureau of Standards, and with specific permission of the Bureau of Mines we will supply the additional information.

II.   SPECTRUM MEASUREMENTS

A.   Noise Measurement Techniques

1.   Description of Measurement Instrumentation

Figure 1 shows a block diagram of the portable portion of the spectrum measuring equipment. Systems 2 and 3 are identical to System 1, so only one system is described. The

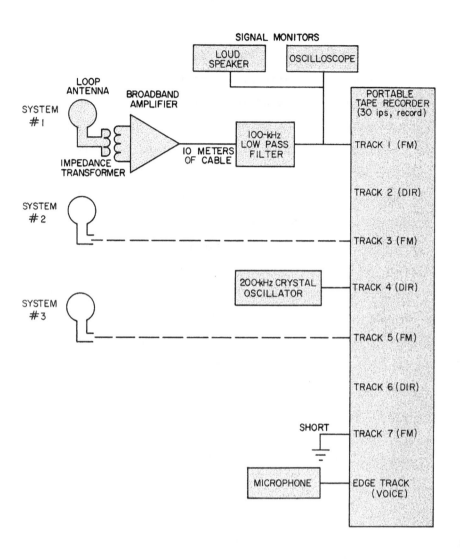

Figure 1   Block diagram of portable instrumentation. FM
            tracks are used to record from 100 Hz to 100 kHz;
            direct tracks are used from 3 kHz to 320 kHz.
            Systems 2 and 3 are identical to system 1. When
            the direct tracks are used, the 100-kHz low pass
            filters are eliminated, and the amplifier band-
            width is increased from 100 kHz to 300 kHz. The
            microphone is used for occasional vocal comments
            by the operator.

6

receiving antenna used for all measurements is a commercially available, electrostatically shielded, 11-turn, 78-cm diameter, air-core loop. A short (40 cm), twin-center-conductor cable connects the antenna to a transformer which steps up the impedance of the antenna to match more nearly the input impedance of the receiver. In order to retain transformer response to about 400 kHz, a compromise between high step-up ratio and broad bandwidth was required. An impedance step-up ratio of 1 to 50 was chosen, giving a voltage step-up of $\sqrt{50}$ or 1 to 7.07. Since the transformer output drives a high-gain receiver input, the transformer is shielded. Shielding consists of a 1/4-inch thick copper enclosure with a layer of high permeability alloy inside the enclosure and a second layer outside the enclosure. In addition to serving as a transformer, this portion of the system contains two other functions. To preclude receiver saturation, a 30-dB attenuator that can be switched in or out is included for use with very high field strengths. Also, an unbalanced input is provided for injecting a field calibration signal (1000-Hz square wave) to assure the system is operating properly when set up in the field. The step-up transformer is connected directly to the balanced inputs of the receiver.

A commercially available, broadband, battery-powered amplifier is used as a receiver. The input impedance is 100 megohms in parallel with 15 picofarads, and the output impedance is 600 ohms. The gain is switch selectable from 10 to 10,000 in 1-2-5 sequence using 1 percent resistors. The bandwidth is also switch selectable, and the passband usually is adjusted between 300 Hz and 100,000 Hz with a 6-dB/octave roll-off outside the pass band for most recordings. For broad-band recordings, the upper bandwidth limit is increased to 300,000 Hz.

The amplifier noise is negligible compared to other sources of system noise at gain settings below 5000. The gain is selected to provide optimum signal level to the analog tape recorder (1.4 volts peak). Ten meters of well-shielded (RG-55) coaxial cable connect the receiver to a low-pass filter. A 100-kHz, low-pass, π filter is used to greatly attenuate received signals at frequencies above the 100-kHz limit of the portable tape recorder.

Two types of signal monitors were used in the mine to determine the general character of the noise fields and to assure that the received, bandlimited signal does not exceed the amplitude capabilities of the tape recorder.

A battery-powered oscilloscope housed in an explosion-proof enclosure is the primary monitoring instrument. By observing the oscilloscope, the instantaneous, peak-to-peak amplitude of all types of noise can be determined readily, and the gain of the amplifier can be appropriately adjusted.

Also, a battery-powered audio amplifier driving a small loudspeaker provides audio monitoring of noise in the audio

range. This device is particularly valuable in catching
transient noise phenomena. During the time spent at a particular recording location, it is impossible to monitor the
oscilloscope visually at all times. For example, during lunch
breaks, while changing tape, and when moving the antenna, the
audio monitor is a useful "alarm" to alert the operator to
unusual, otherwise likely undetected events. Once warned of
these events, it is easier to catch them on tape.

The noise signals processed through the receiving portion
of the portable equipment are recorded on the portable, battery-
powered, analog tape recorder. The frequency modulation (FM)
mode bandwidth is dc to 100 kHz, and the direct mode bandwidth
is 3 kHz to 375 kHz. Figure 1 shows the assignment of FM and
direct tracks with the systems used. Emphasis is placed on
FM usage because of lower recorder signal distortion. The
recorder weighs about 14 kg and is encased in a dust-proof
enclosure. This recorder is a specially modified version of
a commercially available portable tape recorder. The first
modification is a placement of spark-suppressing diodes across
all relay and motor leads to bring the recorder within the
requirements of Schedule 2G of the Bureau of Mines [2]. This
makes the recorder legally permissible and allows it to be
used in explosive atmospheres. A second modification is a
specially built, external, permissible battery box using re-
chargeable sealed lead-acid batteries. The batteries, a solid-

state current limiter, and a fuse, all in series, are enclosed in an explosion-proof enclosure (per Schedule 2G). A single battery allows about four hours of recording. The final modification is the addition of a self-contained 200-kHz crystal oscillator. The output of this oscillator is recorded on track 4 for later use during playback as a reference signal for controlling the speed of the tape-controlled servo in the laboratory analog tape recorder. This is the first method applied to remove flutter, time base error, and sideband generation. Corbin [3] shows a reduction in sideband generation of about 20 dB using this method.

A second method of reducing flutter-sideband generation while using FM mode is to short the input of one channel (channel 7). This signal is retained through the transcription process and is finally inverted and subtracted from all FM signal channels during the digitizing process, as will be described later. This tape recorder is carefully operated and maintained, as it is the most significant source of system noise and distortion. Specially spooled, low-noise tape with 700 m (2300 ft) of tape on a reel is used to increase the recording time from about 11 minutes to about 14 minutes per reel at 30 inches per second (ips, 1 inch = 2.54 cm). Precautions are taken to prevent tape stretching and to reduce print-through. As recommended by Burnett, Corliss, and Berendt [4] all tape recordings are retained on the take-up reel. In addition to better tape packing and less print-through, leaving the tape on the take-up reel precludes the time-consuming process of rewinding while on site in the mine.

The portable equipment has a total weight of about 40 kg. Removal of the permissibility requirement would reduce the weight by approximately one-third.

2. Description of Processing Instrumentation

a. Laboratory Transcription Process

After the mine visit, the newly recorded tapes are reviewed in the laboratory at the record speed of 30 ips to note any pertinent recorded vocal comments and to select the data to be processed.

For the initial step of processing, the portable tape recorder is reconfigured to play back at a tape speed of half the record speed (15 ips). This is the initial step of frequency range reduction performed to reduce the data bandwidth ultimately to the requirement of 5 kHz imposed by the digitizer. In this process, all of the original information, including high frequencies, is conserved. The difference now is that a data signal at 100 kHz lasting, say, 1 second, comes out at 50 kHz and lasts 2 seconds.

Figure 2 shows the block diagram of the laboratory transcription process. Only tracks 2 and 4 of the laboratory tape recorder are instrumented to process direct mode recordings. If both tracks 2 and 6 on the portable recorder contain direct

11

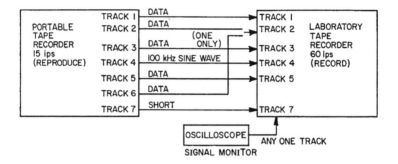

Figure 2   Block diagram of laboratory transcription process.
This is the first step in frequency range reduction. Newly recorded tape is played back at one-
half speed.

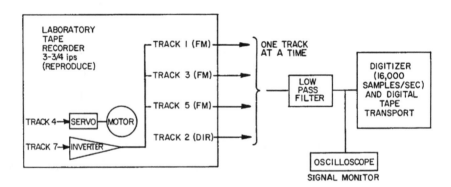

Figure 3   Block diagram of laboratory digitizing of data.
Track 7 inverter output is added to the outputs of
tracks 1, 3 and 5 to reduce distortion. This is
the final step of frequency range reduction.

12

data, then two runs are made. Note that the crystal oscillator reference signal (track 4 on the portable recorder) is now 100 kHz. Operating the laboratory tape recorder at 60 ips and recording a servo frequency of 100 kHz meets the Inter-Range Instrumentation Group (IRIG) standard (according to the manufacturer) for which the manufacturer configured the laboratory recorder. As with the portable recorder, manufacturer-recommended low-noise tape, tape storage on take up reels, and other precautions are taken with the laboratory tape recorder.

b. Laboratory Digitizing Process

The laboratory digitizing of data is shown in Figure 3. The desired frequency range of spectral display must be considered in this step. The considerations of tape recorder speed versus aliasing, predigitizer filter cut-off frequency selection, digitizer rate, etc., will not be discussed here. In general, the laboratory tape recorder is run at 1-7/8 ips for an upper frequency display limit of 320 kHz (direct mode), and 3-3/4 ips for 100 kHz and 3 kHz (FM mode). The most frequently used analysis uses a recorder playback speed of 3-3/4 ips for a frequency display limit of 100 kHz. During this step the laboratory recorder speed-control servo (including the motor) is controlled by the crystal oscillator signal originally recorded at 200 kHz in the field. Most of the

13

accumulated flutter, wow, time base error, and sideband generation is eliminated. Also, the data are reduced to the 5-kHz frequency range imposed by the 16,000-sample-per-second maximum sample rate of the digitizer.

c. Spectral Computation and Graphical Output

A 12-bit conversion system digitizes the data and records the converted data onto digital magnetic tape for subsequent computer processing. A study using 6-, 8-, 10-, and 12-bit data determined the number of bits required. The study indicated that at least 10 bits, and usually 12 bits, are required to provide enough dynamic range satisfactorily, especially for the narrow range (750 Hz) analysis.

As outlined in figure 4, a large digital computer is used to compute the spectra and to print the results on an integral high-speed cathode-ray tube microfilm plotter. The individual microfilm frames are subsequently joined to give the spectra found in this report.

d. Description of Processing Software

Several programs are used in the processing of the digit-ized data. They are listed here for reference purposes only without further elaboration: PSPECSET, PSPECZRO, PSPECINP, PSPECEST, PSPECOUT, PSPECCAP, SPEAKS, FILMGRAF, and DDGRAPH. These, and several other subroutines, were developed primarily

14

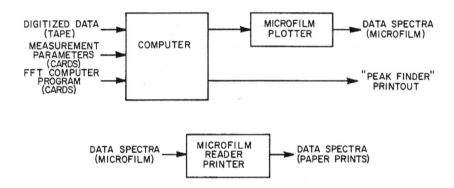

Figure 4   Block diagram of spectral computation, plotting, and printing.

by L.D. Lewis of the Space Environment Laboratory of the National Oceanic and Atmospheric Administration, Boulder, Colorado 80302.

A calling program identified as ELF was written to use and adapt the software listed above to the needs of this project. It also serves to adapt the output from the analog-to-digital converter to a computer-usable format.

It is outside the scope of this report to describe in any detail the theory of the Fast Fourier Transform (FFT) or the internal organization of the software. The program package is based on a paper by Welch [5], and more information is available in his paper and references.

A summary of the processing parameters for the most frequently used bandwidth display (100 kHz) follows:

1. Each printed spectrum (other than 3-D) is an average of twenty spectra.

2. The ensemble length is 2^{12}, i.e., 4096, 12-bit words. This is equal to 0.008 seconds of real time. For the 50 percent data overlap factor used, (i.e., the first spectrum used samples number 1 through 4096, the second used 2048 through 6144, etc.) each average of 20 spectra represents 0.08 seconds of real time.

3. Spectral resolution is 125 Hz over the 100,000 Hz frequency range.

16

4.   Cost of computer time per spectrum is about $0.15.
     The cost of a plot of 20 averaged spectra is
     therefore about $3.00. The 3-D plots use different
     parameters, display hundreds of averaged spectra,
     and cost about $25.00 each for computer time.

An important improvement achieved in our software is the
virtual elimination of an inherent potential 3-dB error in the
reported amplitude of cw signals. This error arises from the
uncertainty of the frequency of a cw signal, and therefore of
its location within any single 125-Hz spectral resolution
analysis bandwidth. For example, if the cw signal falls on the
dividing line between 125-Hz spectral resolution "cells", its
power would be reported to be 3 dB (1/2 power) below that reported if it were in the center of a cell. By using the rule
that the area (power) under a spectral density curve represent-
ing a cw signal is constant (Parsival's Theorem), a correction
to the height of cw signals is applied. This correction has
been demonstrated to give a reduction of the 3-dB maximum
error to less than 0.05 dB. We have named this program "Peak
Finder".

The processing software also includes a computer routine
that makes the system present a uniform frequency response
over the bandwidth of interest. This software, a type
of correction curve algorithm, will be discussed under antenna
calibration.

17

3. System Evaluation

In order to provide processed data referenceable to absolute field-strength standards, it was necessary to measure
the overall system response and establish that certain error
sources were acceptably small. The system parameters measured
included: receiver gain linearity, dynamic range, system gain
drift, frequency accuracy, and harmonic and intermodulation
distortion.

The switchable gain settings (providing an amplifier gain
of 10 to 10,000) on the three receiving amplifiers were checked
at 450 Hz and 40 kHz. All amplifier switch settings are
linear in gain within 0.1 dB with one exception. System
number 1 (used mostly for vertical antenna sensitive axis
measurements) gain is 0.8 dB low at a gain setting of 10,000
using the test frequency of 40 kHz.

Instantaneous dynamic range was measured by injecting a
sine wave at the highest amplitude the recorder can tolerate
(1 Vrms) and then reducing the amplitude with a step attenuator
until the signal disappeared into the system noise. Instantaneous dynamic range for the 1-to-100 kHz spectra (FM recording technique) is 62 dB using a test signal frequency of
40 kHz. Higher instantaneous dynamic ranges are available for
lower bandwidth spectra, e.g., about 90 dB for 750 Hz spectra.

The above instantaneous dynamic range tests were made with
a receiver gain of 100. With a receiver gain of 10,000, signals

18

30 dB lower can be measured. Higher amplitude signals can be measured by reducing the receiver gain to 10, and very high amplitude signals can be measured by switching in the 30 dB attenuator in the antenna transformer and balun network. The total of the instantaneous dynamic range and switchable gain is therefore 170 dB (90 + 30 + 20 + 30). This is the range of signal amplitude that can be measured with this system for spectra with a 750 Hz bandwidth.

The system gain stability was measured by calibrating each of the three systems against a standard H field three times over a six-month period. Using 20 comparable measurements (i.e., the same system with the same gain, at the same frequency) made two months apart, the mean system gain increased 0.56 dB with a standard deviation of 0.25 dB. Using another set of 30 comparable measurements made four months apart (not overlapping the above two-month period), the mean system gain increased 0.44 dB with a standard deviation of 0.21 dB. The highest single gain increase was 1.37 dB for system 2 over the two-month interval at 100 kHz. This occured at the highest frequency in a particular band, and was due largely to variations of filter characteristics which are dominant at the high end of the pass band.

The frequency accuracy was measured by reading the frequency of WWVB (60,000 Hz) as reported on the "peak-finder" program computer printout. ("Peak-finder" is the program that corrects for the minus 3 dB uncertainty mentioned in an earlier paragraph, and also determines frequency). The average frequency from 110 measurements made over a six-month period was 59,986.9 Hz (0.022% low) with a standard deviation of 4.3 Hz (0.0072%). (The frequency offset is probably due to offsets in the crystal oscillators in the tape recorders or at the digitizer.) Thus, after adding a 0.022% correction, frequencies can be measured with a standard deviation of 0.0072% (at least at 60 kHz).

Harmonic distortion was measured by injecting a 40 kHz sinewave. For an input level 10 dB below the full-rated input of the tape recorder (1 Vrms), the second harmonic is down 44 dB. For an input of 1 Vrms, the second harmonic is down 35 dB.

Intermodulation distortion on the 1-to-100 kHz spectrum was measured by injecting simultaneously two sine waves of equal amplitude at frequencies of 45 kHz and 55 kHz. The sum and difference frequencies are down 49 and 46 dB respectively.

System bandwidths and digitizer rates were chosen to attenuate aliased signals by 60 dB or more. A test 5685 Hz square wave was injected to test for aliasing. No evidence of aliasing was found, on this test, nor during the course of our measurements.

The most significant source of error found to date is the inaccuracy of the mathematical fit of the system gain correction to the known system response. For the chosen level

20

of effort expended on fitting functions, the largest error produced by inaccurate fit is estimated to be 1.0 dB.

Much of the spectra above 10 kHz in Robena is of such low noise level as to be obscured by system noise. Subsequent equipment modifications have allowed selected channels to attenuate frequencies below 10 kHz so that the 62 dB dynamic range of the wideband (1 kHz to 100 kHz) system can be used more effectively. With strong noise components below 10 kHz attenuated, higher system gain (and therefore lower system noise as referred to the front end) can be achieved.

Reduction of the system noise level is readily obtainable for reduced bandwidth recordings (e.g., 100 Hz to 10 kHz). Antenna transformer (balun) design dictates a trade off between step-up ratio and bandwidth. The transformer with lower step-up ratio but wider bandwidths was selected for use in Robena.

Further reduction of the system noise would be possible by eliminating the analog tape recorder and substituting a method of high-speed portable digital recording.

It should be emphasized, however, that the noise level of the recording system is of such low value that the noise data presented should be satisfactory for many or most system design studies. In cases where the measured noise was equal to or below the system noise, the system noise floor established an upper limit which the mine noise did not exceed.

21

4. System Calibration

The loop antenna calibration is performed by applying a
known magnetic field to the receiving loop antenna at the
National Bureau of Standards (NBS) loop calibration facility
located in Boulder, Colorado. The field generated at this
facility has a reported uncertainty of ± 3 percent (± 0.26
dB) [6] over the frequency ranges used.

Strictly speaking, the calibration is dependent on the
surroundings, but the dependence is slight, since the loop is
small compared to a wavelength in all the nearby media. The
highest frequency measured is the fourth harmonic of 88 kHz,
i.e., 352 kHz, with a wavelength of 852 meters. From Faraday's
law, the loop measures the time derivative of the component
of magnetic induction B normal to the plane of the loop,
integrated over the area of the loop. The mathematical definition of H, and a more comprehensive description of the calibration site are given on page 94.

The antenna is not calibrated as an independent component; all instrumentation is calibrated as a system by
using the reduced data from the microfilm plotter as the output indicator. Thus, the system performance, including that
of the software, is measured. The resulting gain corrections
are applied directly to the raw spectra to produce spectra
that are directly readable as absolute field strength.

The current probe used in making the measurements is
calibrated by clamping it around a wire carrying a known

22

current. The transmitting standard loop antenna was used as the wire with known current. Therefore, the uncertainty of ± 3 percent (0.26 dB) reported for the field [6] also applies to the current used to generate the field. As with the loop antenna, the entire system is calibrated by using the microfilm plotter as the output indicator.

The voltage probe used in making the measurements is calibrated by injecting a known sinusoid into the probe, again using the microfilm plotter as the output indicator. The amplitude of the sinusoid is adjusted to a constant value at each frequency using a commercially available rms voltmeter. The voltmeter accuracy is advertised to be ± 1 percent (0.086 dB) of full scale. The rms voltmeter was checked on a commercial laboratory voltmeter calibrator (with an advertised accuracy of ± 0.2 percent) and found to be within specifications.

The total system uncertainty is a composite of calibration field uncertainties and system instabilities. Each has been discussed, and the user may combine the uncertainties as he deems best. A total uncertainty of ± 1 dB is felt to be conservative in most cases.

B. Noise Measurement Results: Spectra

1. Mine Description and Antenna Sites

Measurements were made in the Robena No. 4 coal mine located near Waynesburg, in southwestern Pennsylvania. Figure 5 shows a map of working section 3-main, 10-right, 2-room near

23

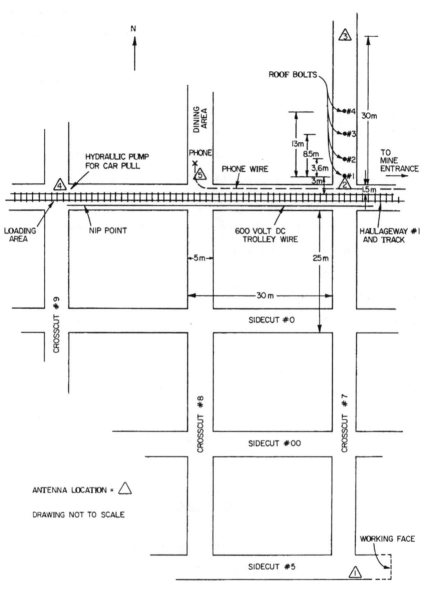

Figure 5   Map of working section 3-main, 10-right, 2-room, in Robena No. 4 coal mine as it was on December 5, 1972 when measurements were being made.

Blaker Shaft where the majority of measurements were made on December 5 and 7, 1972. The overburden in this area varies between 200 and 300 meters. The entire mine, including all machinery, is powered by 600 volts dc. A combination of igni-tron rectifiers and rotary converters is used. All conversion from ac to dc is done on the surface with the result that no ac power is brought into this mine.

2. Electromagnetic Noise Spectrum Results

a. Introduction

When reading values from the 100-kHz spectra in this report, keep the following points in mind:

1. Note the roll-off frequencies. Values above 100 kHz and below 1 kHz are not calibrated. Because of this, do not attempt to read values above 100 kHz or below 1 kHz.

2. The correct units for the spectral peaks are micro-amperes per meter (μA/m), since they are narrower than the spectral resolution of the plots.

3. The broad-band noise between spectral peaks is as seen by a receiver having the same bandwidth as the Fast Fourier Transform (FFT) spectral resolution bandwidth used to compute the spectrum (125 Hz for 1-to-100-kHz graphs). The correct units for the background noise between peaks are microamperes per meter per square root x hertz $[(\mu A/m)/\sqrt{x\ Hz}]$, where x is the spectral resolution of the FFT (x equals 125 Hz for the 1-to-100-kHz graphs).

25

An easy way to obtain the spectral density per (one) root
hertz when reading broad-band noise is to subtract the required
number of dB, remembering that the units have now changed
to $(\mu A/m)/\sqrt{Hz}$. For spectra with a resolution bandwidth of
125 Hz, subtract 20.97 dB, for 62.5 Hz subtract 17.96 dB, and
for 7.81 Hz subtract 8.93 dB.

The Appendix gives the code key used in determining the
meaning of the numbers in the block at the top of each spectrum.
The resolution bandwidth is given on the ordinate of the plots.

b. Working Face Area

Figure 6, upper curve, shows the magnetic field noise
spectrum received at the antenna location identified as 1 (in
figure 5). The lower curve in this, and in following spectra,
is the receiving system noise. It is included to indicate frequency ranges in which system noise may predominate. The lower
curve is obtained by replacing the antenna with a dummy antenna.
In figure 6, mine noise is higher than system noise at all frequencies. Note that the system noise varies with gain when it
is expressed as equivalent input noise. The antenna loop was
placed flat on the ground (the sensitive axis for near fields
was therefore pointed up-down, i.e., vertically). The antenna
position was approximately 10 meters behind the back end of the
advancing continuous mining machine (continuous miner). One of
the two shuttle cars was positioned immediately behind the miner

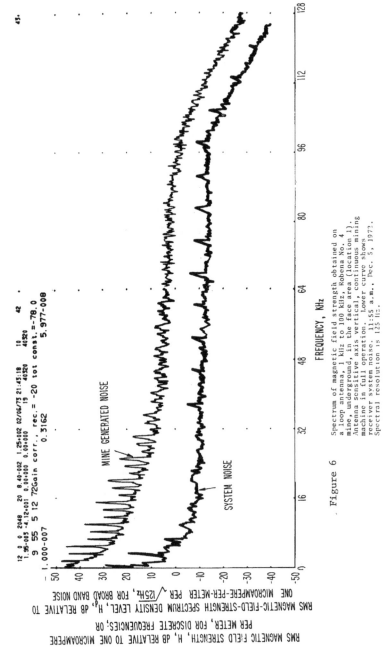

Figure 6

Spectrum of magnetic field strength obtained on
a loop antenna, 1 kHz to 100 kHz. Robena No. 4
mine, underground, in the face area (location 1).
Antenna sensitive axis vertical, continuous mining
machine in full operation. Lower curve shows
receiver system noise. 11:55 a.m., Dec. 5, 1972.
Spectral resolution is 125 Hz.

and was receiving coal. The miner power and water cables were on the floor about 2 meters away from the antenna. Probably a large majority of the energy received by the antenna was radiated by the power cable. The highest spectral peaks of field strength, 48 dB μA/m, 45 dB μA/m, and 46 dB μA/m, etc., at frequencies of 1.67 kHz, 3.33 kHz, 5 kHz, respectively, etc., were due to rotating electric machinery. These peaks were slowly shifting upward in frequency as will be shown later in 3-D plots. The peaks were caused by an electrical circuit being closed and opened 1.67 thousand times per second, producing that number of small impulses per second. Commutator bars rotating under a brush on a dc motor were the probable source.

Figures 7 and 8, upper curves, show noise signatures picked up by two other orthogonal antennas. The loop antennas were standing on edge with the horizontal antenna sensitive axis directed E-W in the first case and N-S in the second case. All three noise signatures shown in figures 6, 7, and 8 were taken simultaneously using the three antennas and three separate tracks on the tape recorder. At least three conclusions can be made about these figures. First, the horizontal sensitive axis noise received is lower by 10 to 30 dB than the vertical sensitive axis noise. Second, the horizontal sensitive axis noise reaches a relative minimum in the region of 32 kHz and then rises 10 dB or so to an apparent maximum in the region of

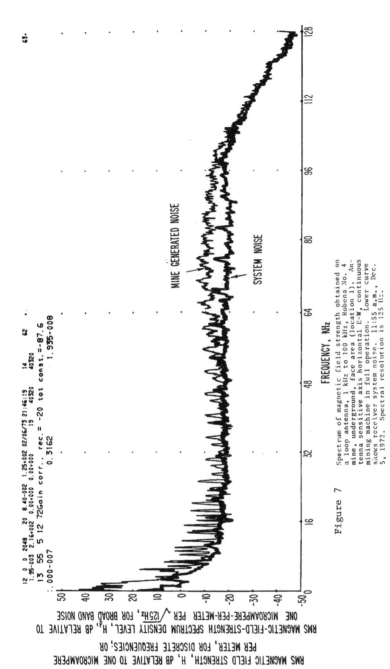

Figure 7

Spectrum of magnetic field strength obtained on a loop antenna, 1 kHz to 100 kHz, Robena No. 4 mine, underground, face area (location 1). Antenna sensitive axis horizontal E-W, continuous mining machine in full operation. Lower curve shows receiver system noise. 11:55 a.m., Dec. 5, 1972. Spectral resolution is 125 Hz.

29

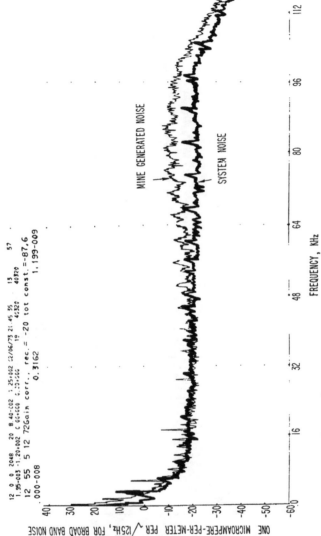

50.

12 0 0 2048   20 8.40+002 -1.25+002 22/06/73 21.45.55   15   57
1.95+003 -1.20+002 C.00+000 C.00+000   19   40320
12 55 5 12 72Gain corr., rec. = -20 tot const. =-87.6
      .000-008   0.3162   1.199-009

MINE GENERATED NOISE

SYSTEM NOISE

FREQUENCY, kHz

RMS MAGNETIC-FIELD-STRENGTH SPECTRUM DENSITY LEVEL, H, $\sqrt{/125Hz}$, dB RELATIVE TO
ONE MICROAMPERE-PER-METER PER $\sqrt{/125Hz}$, FOR BROAD BAND NOISE

RMS MAGNETIC FIELD STRENGTH, H, dB RELATIVE TO ONE MICROAMPERE
PER METER, FOR DISCRETE FREQUENCIES, OR

Figure 8

Spectrum of magnetic field strength obtained on
a loop antenna, 1 kHz to 100 kHz, Robena No. 4
mine, underground, face area (location 1). Antenna sensitive axis horizontal N-S, continuous
mining machine in full operation. Lower curve
shows system noise, 11:55 a.m., Dec. 5, 1972.
Spectral resolution is 125 Hz.

30

80 kHz. These are the only two spectra taken in Robena No. 4
that show this broad rise. Possibly it is necessary to be
close to the noise source to observe this. A third observation
is the extension of spectral features, 5-kHz wide, up to 100
kHz. These features are shown to be shifting upward in frequency with time in the 3-D display shown in figure 9. The
machine producing the noise was increasing its mechanical speed
during the time covered, thereby causing the frequency of output noise to increase. Figure 9 displays a frequency range of
21 kHz, and there are four visible repetitions of the feature
that is about 5-kHz wide. Since these features move in frequency with time, the peaks of the noise at any given time
could appear at any frequency. Therefore, when considering
reading noise values at a particular frequency, the highest
adjacent values probably should be the ones used. Keep in
mind also that the noise presented here is continuous in
nature. Short-duration impulses were observed earlier that
exceeded the noise levels presented in the plots so far. Some
typical examples of impulses will be given later.

Figures 10, 11, and 12 are the same spectra as shown in
figures 6, 7, and 8, but the frequency scale is expanded.
These spectra have a 7.81 Hz resolution bandwidth. Field
strengths can be read from these spectra in the range from
100 Hz to 3 kHz; outside this range they are not calibrated.
Figure 10 shows that there is some broadband noise present be-

31

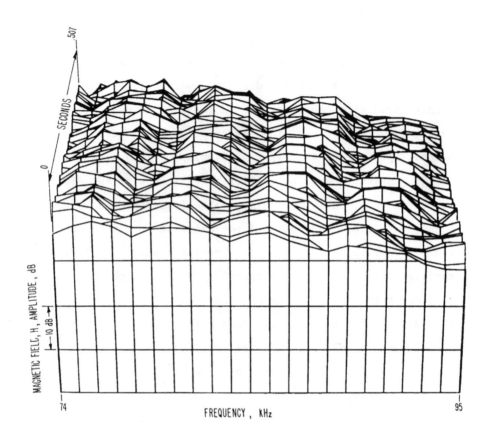

Figure 9   Three-D spectrum of magnetic field strength ob-
            tained on a loop antenna, 74 kHz to 95 kHz, Robena
            No. 4 mine, face area. Time progression presenta-
            tion showing continuous mining machine noise in-
            creasing in frequency as a function of time.
            Spectral resolution is 1 kHz. Relative amplitude
            is shown.

32

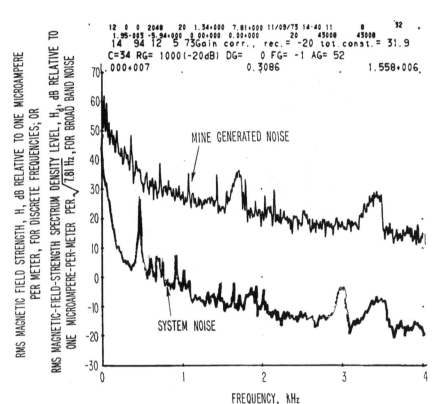

Figure 10   Spectrum of magnetic field strength obtained on
a loop antenna, 100 Hz to 3 kHz, Robena No. 4
mine, underground, face area (location 1). Antenna sensitive axis vertical. Continuous
mining machine in full operation. Lower curve
shows receiver system noise. 11:55 a.m., Dec.
5, 1972. Spectral resolution is 7.81 Hz.

33

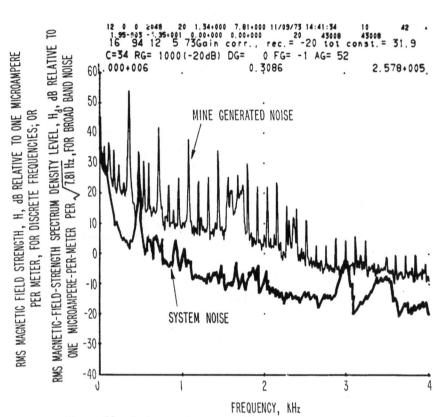

Figure 11   Spectrum of magnetic field strength obtained on
a loop antenna 100 Hz to 3 kHz, Robena No. 4
mine, underground, face area (location 1). Antenna sensitive axis horizontal E-W. Continuous
mining machine in full operation. Lower curve
shows receiver system noise. 11:55 a.m., Dec.
5, 1972. Spectral resolution is 7.81 Hz.

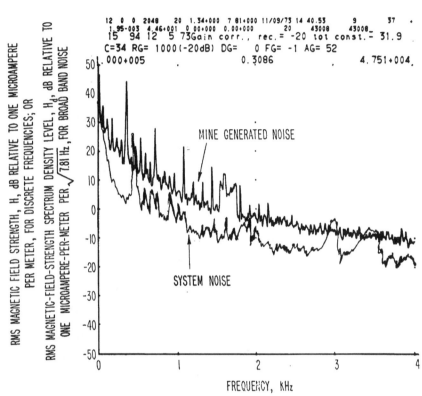

Figure 12   Spectrum of magnetic field strength obtained on a loop antenna 100 Hz to 3 kHz, Robena No. 4 mine, underground, face area (location 1). Antenna sensitive axis horizontal N-S. Continuous mining machine in full operation. Lower curve shows receiver system noise. 11:55 a.m., Dec. 5, 1972. Spectral resolution is 7.81 Hz.

tween the commutator-generated noise peaks. Figures 11 and 12
show that even with face machinery in operation, powerline
harmonics are predominant in this frequency range for horizontal antenna sensitive axis orientation.

Figure 13 shows the noise received at the face area during
lunch hour (quiet time) and in the absence of any electrical
machinery (the miner was parked elsewhere). In this case the
system noise (primarily the tape recorder) limits sensitivity
to about -20 dB relative to 1 μA/m. Figure 14 shows the same
spectrum as Figure 13, but with the frequency scale expanded.
The first three powerline harmonics (of 360 Hz) were stronger
during lunch hour than when the equipment was present and
working. In the region between 1080 and 2880 Hz, the 3rd
through 8th harmonics of 360 Hz are found. The 360-Hz noise
is generated when 3-phase 60-Hz ac power is rectified to provide mine operating dc power. Technically, without any filter-
ing, the mine power is pulsating dc (pulsating at 360 pulses
per second). The levels of the harmonics are 41 dB μA/m for
the 3rd harmonic, 36 dB for the 4th, 32 dB for the 5th, 25 dB
for the 6th, 12 dB for the 7th, and 9 dB for the 8th. The loga
rithmic average (average dB value) of these six harmonics is
26 dB μA/m. These values are obtained from a computer printout from the program "Peak Finder".

36

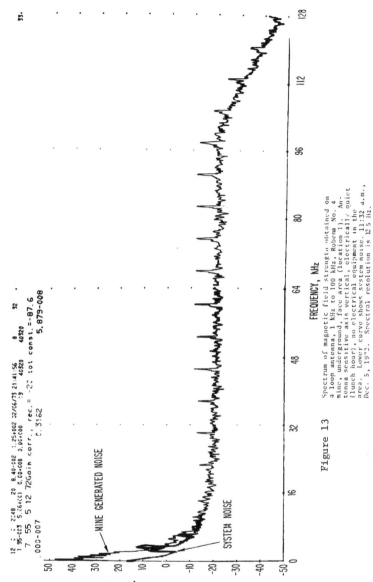

Figure 13

Spectrum of magnetic field strength obtained on a loop antenna, 1 kHz to 100 kHz, Robena No. 4 mine underground, face area (location 1). Antenna sensitive axis vertical, electrically quiet (lunch hour), no electrical equipment in the area. Lower curve shows system noise. 11:32 a.m., Dec. 5, 1972. Spectral resolution is 125 Hz.

RMS MAGNETIC-FIELD-STRENGTH SPECTRUM DENSITY LEVEL, H_d, dB RELATIVE TO ONE MICROAMPERE-PER-METER PER $\sqrt{125Hz}$, FOR BROAD BAND NOISE

PER METER, FOR DISCRETE FREQUENCIES, OR

RMS MAGNETIC FIELD STRENGTH, H, dB RELATIVE TO ONE MICROAMPERE

37

c. At Rail of Haulageway (Location 2)

Figures 15, 16, and 17 show the three orthogonal components of noise recorded at antenna location 2, which is 1.5 m
from the nearest rail in the main haulageway (refer to figure 5
for antenna location). The time is 2:35 p.m., between shifts,
with a medium-sized (13 ton) locomotive arriving pulling a
man-trip car with the second shift complement of workers.
Apparent on these three plots are the higher levels of 360 Hz
and associated harmonics next to the trolley, as compared to
their levels on figure 10 during lunch hour at the face. Figures 18, 19, and 20 show expanded spectra of this event. From
figure 20 (antenna sensitive axis pointing towards trolley
wire) the harmonic magnetic field strengths are 73 dB μA/m
for the 3rd harmonic, 69 dB for the 4th, 65 dB for the 5th,
59 dB for the 6th, 48 dB for the 7th, and 42 dB for the 8th.
The logarithmic average strength of harmonics 3 through 8 is
59 dB μA/m. The logarithmic average strength of harmonics
3 through 8 at the face (vertical moment) was 26 dB μA/m.
Subtracting the two averages, we find that field strengths
in the haulage way, between 1080 Hz and 2880 Hz, on a
logarithmic basis, are higher by 33 dB, a power ratio of
about 2000 to 1. Later we will show evidence of much higher
field strengths near main dc power feed lines.

Figures 21, 22, and 23 are 3-D plots showing how the
noise changes as a function of time. The first portions of

38

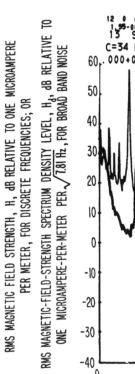
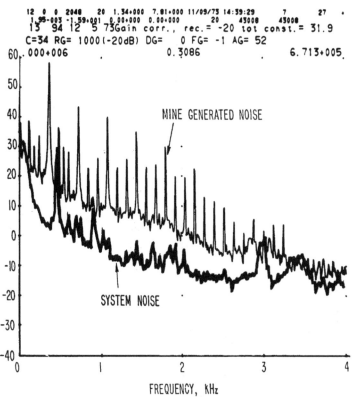

RMS MAGNETIC FIELD STRENGTH, H, dB RELATIVE TO ONE MICROAMPERE PER METER, FOR DISCRETE FREQUENCIES; OR

RMS MAGNETIC-FIELD-STRENGTH SPECTRUM DENSITY LEVEL, H_d, dB RELATIVE TO ONE MICROAMPERE-PER-METER PER $\sqrt{7.81\,Hz}$, FOR BROAD BAND NOISE

MINE GENERATED NOISE

SYSTEM NOISE

FREQUENCY, kHz

Figure 14   Spectrum of magnetic field strength obtained on a loop antenna 100 Hz to 3 kHz, Robena No. 4 mine, underground, face area (location 1). Antenna sensitive axis vertical. Quiet (lunch hour), no electrical equipment in the area. Lower curve snows system noise. 11:32 a.m., Dec. 5, 1972. Spectral resolution is 7.81 Hz.

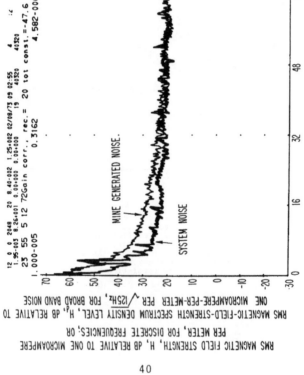

Figure 15

Spectrum of magnetic field strength obtained on a
loop antenna 1 kHz to 100 kHz, Robena No. 4 mine,
underground, crosscut No. 7, antenna sensitive
axis vertical, 1.5 m from track, 2:35 p.m., loc.
5, 1972. Thirteen-ton locomotive pulling miners
into section. Spectral resolution is 125 Hz.

FREQUENCY, KHz

RMS MAGNETIC-FIELD-STRENGTH SPECTRUM DENSITY LEVEL, H₁, dB RELATIVE TO
ONE MICROAMPERE-PER-METER PER √125Hz, FOR BROAD BAND NOISE

RMS MAGNETIC FIELD STRENGTH, H, dB RELATIVE TO ONE MICROAMPERE
PER METER, FOR DISCRETE FREQUENCIES; OR

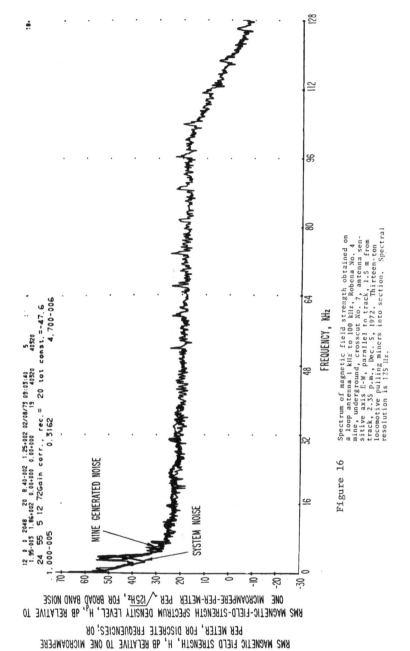

Figure 16   Spectrum of magnetic field strength obtained on a loop antenna 1 kHz to 100 kHz, Robena No. 4 mine, underground, crosscut No. 7, antenna sensitive axis E-W, parallel to track, 1.5 m from track, 2.35 p.m., Dec. 5, 1972. Thirteen-ton locomotive pulling miners into section. Spectral resolution is 125 Hz.

41

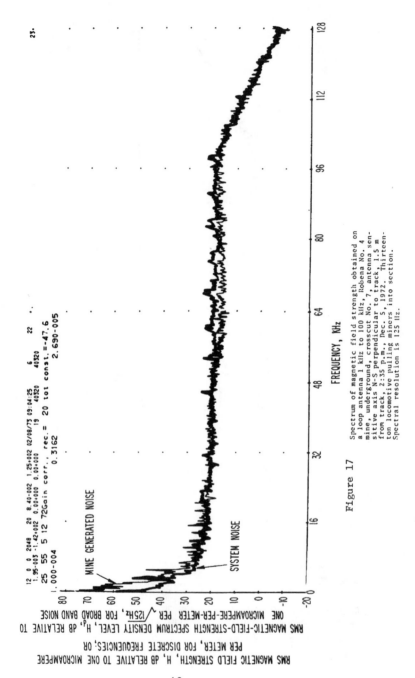

Figure 17

Spectrum of magnetic field strength obtained on a loop antenna 1 kHz to 100 kHz, Robena No. 4 mine, underground, crosscut No. 7, antenna sensitive axis N-S perpendicular to track 1.5 m from track, 2:35 p.m., Dec. 5, 1972. Thirteen-ton locomotive pulling miners into section. Spectral resolution is 125 Hz.

42

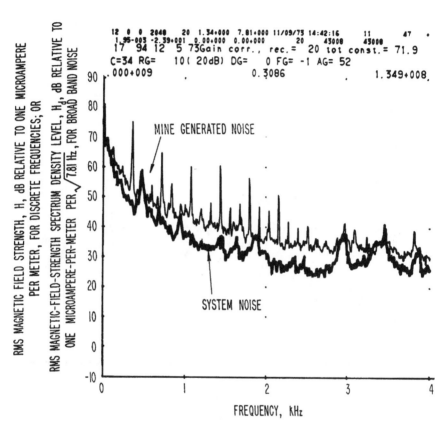

Figure 18　Spectrum of magnetic field strength obtained on a loop antenna 100 Hz to 3 kHz, Robena No. 4 mine, underground, crosscut No. 7, 1.5 m from rail, antenna sensitive axis vertical, 2:35 p.m., Dec. 5, 1972. Thirteen-ton locomotive pulling miners into section. Spectral resolution is 7.81 Hz.

43

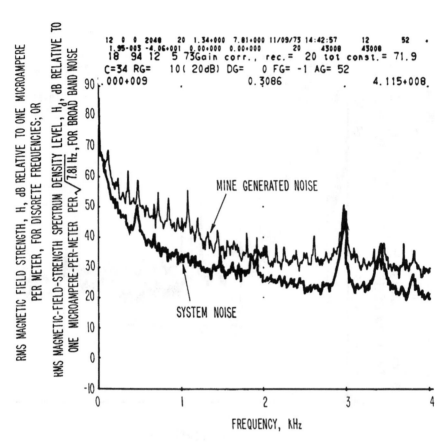

Figure 19   Spectrum of magnetic field strength obtained on
a loop antenna 100 Hz to 3 kHz, Robena No. 4
mine, underground, crosscut No. 7, 1.5 m from
rail, antenna sensitive axis horizontal E-W,
2:35 p.m., Dec. 5, 1972. Thirteen-ton locomotive pulling miners into section. Spectral
resolution is 7.81 Hz.

44

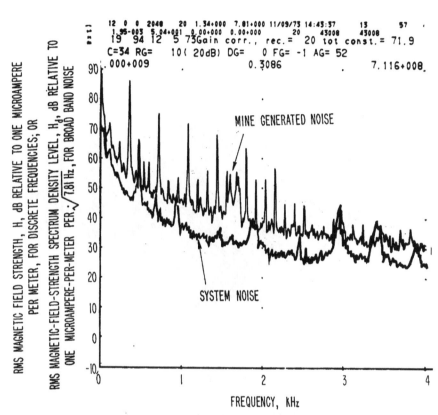

Figure 20   Spectrum of magnetic field strength obtained on
a loop antenna 100 Hz to 3 kHz, Robena No. 4
mine, underground, crosscut No. 7, 1.5 m from
rail, antenna sensitive axis horizontal N-S,
2:35 p.m., Dec. 5, 1972. Thirteen-ton locomotive pulling miners into section. Spectral
resolution is 7.81 Hz.

45

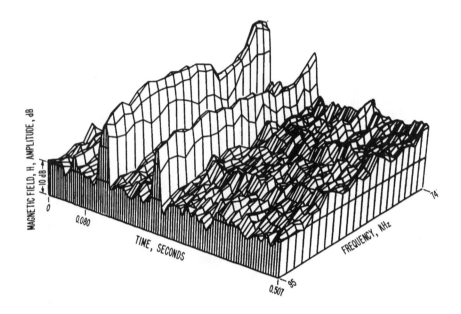

Figure 21   Three-D spectrum of magnetic field strength ob-
             tained on a loop antenna 74 kHz to 95 kHz, Robena
             No. 4 mine, underground, crosscut No. 7, loop
             antenna, antenna sensitive axis vertical, 1.5
             meters from nearest rail. Thirteen-ton locomo-
             tive pulling miners into section. Spectral reso-
             lution is 1 kHz. Relative amplitude is shown.

46

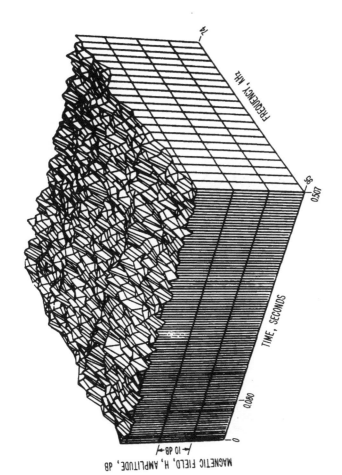

Figure 22   Three-D spectrum of magnetic field strength obtained on a loop antenna 74 kHz to 95 kHz, Robena
No. 4 mine, underground, crosscut No. 7, antenna
sensitive axis horizontal E-W, pointed parallel to
haulage way, 1.5 meters from nearest rail.
Thirteen-ton locomotive pulling miners into section. Spectral resolution is 1 kHz. Relative
amplitude is shown.

47

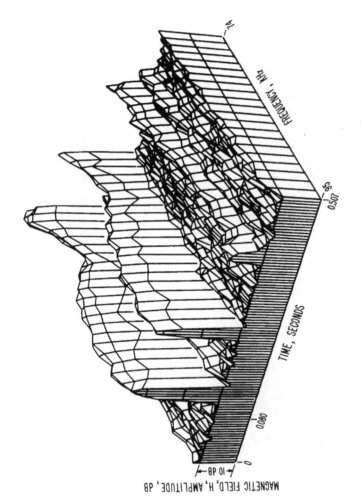

MAGNETIC FIELD, H, AMPLITUDE, dB

Figure 23   Three-D spectrum of magnetic field strength obtained on a loop antenna 74 kHz to 95 kHz, Robena
No. 4 mine, crosscut No. 7, loop antenna, antenna
sensitive axis horizontal N-S pointed toward
trolley wire, 1.5 meters from nearest rail.
Thirteen-ton locomotive pulling miners into section.   Spectral resolution is 1 kHz.   Relative
amplitude is shown.

these plots were shown in the six preceding spectra (antenna location 2). Here, the frequency region between 74 kHz and 95 kHz is selected for analysis over about a 1/2-second period with 1-kHz spectral resolution. Immediately apparent on two of these graphs are impulses of varying strength. These impulses are caused by the approaching 13-ton locomotive. The strongest impulses are again received by the antenna with horizontal sensitive axis oriented N-S (pointed toward the trolley wire). This is not too surprising, as the trolley wire and rails probably act to transmit the noise to some extent. The vertical scale on the 3-D plots is incremented in 10-dB steps. The height of the strongest impulse is about 25 dB above the noise floor. The noise floor here represents the receiver noise, and for the gain settings used for this measurement this floor is about 20 dB above 1 µA/m. The impulse can then be inferred to be about 45 dB µA/m. These pulses are typical and should not be considered as maximum field strengths encountered. The APD information given elsewhere in this report will give more accurate information on distributions of pulse amplitude for the measurement bandwidth. Note that for each 3-D graph, the information shown during the interval of time, 0 to 80 ms, is averaged for presentation in the corresponding 2-D spectra in figures 15, 16, and 17.

These three figures (21, 22, 23) also illustrate the necessity of measuring three orthogonal components simultaneously to obtain the field at any given point in the mine. For example, if only one horizontal component (fig. 22) were measured, the impulses present at this point (as shown in figures 21 and 23) would have been entirely missed.

Other measurements made at antenna location number 2, including roof-bolt voltage, trolley-wire voltage, and phone line current and voltage, will be discussed later in section IV, Special Measurements.

      d.   Thirty Meters from Rail (Location 3)

Antenna location number 3 is 31.5 meters from the track in crosscut number 7, or 30 meters farther away from the track than location number 2. Figures 24, 25, and 26 show the three orthogonal components of noise received at location number 3. All three orthogonal components generally show 20 to 30 dB less noise than received at location 2, in the frequency region below 10 kHz. The 88-kHz trolley phone shows up clearly at 32 dB µA/m for the vertical sensitive axis, and 25 dB µA/m and 23 dB µA/m for horizontal sensitive axes E-W and N-S, respectively. The trolley phone signal received on a vertical sensitive axis antenna is 7 and 9 dB higher than the signals received on the two horizontal sensitive axis antennas.

Figures 27, 28, and 29 show expanded spectra for the above location 3. Power-line harmonics are strongest on the antenna with the sensitive axis horizontal N-S, pointing toward the trolley wire and tracks.

50

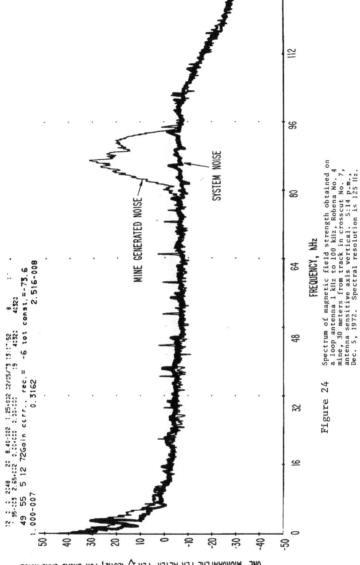

Figure 24   Spectrum of magnetic field strength obtained on
a loop antenna 1 kHz to 100 kHz, Robena No. 4
mine, 30 meters from track in crosscut No. 7,
antenna sensitive axis vertical. 5:14 p.m.,
Dec. 5, 1972. Spectral resolution is 125 Hz.

51

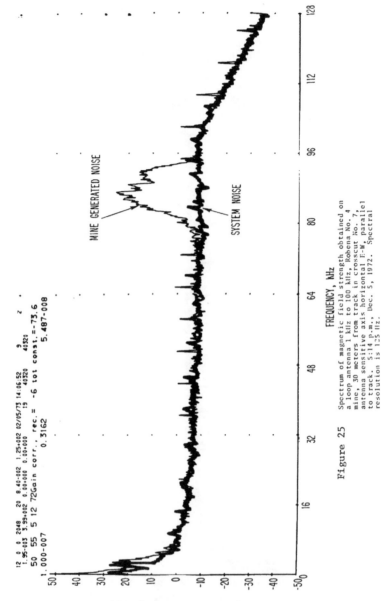

Figure 25   Spectrum of magnetic field strength obtained on
a loop antenna 1 kHz to 100 kHz, Robena No. 4
mine, 30 meters from track in crosscut No. 7,
antenna sensitive axis horizontal E-W, parallel
to track, 5:14 p.m., Dec. 5, 1972. Spectral
resolution is 125 Hz.

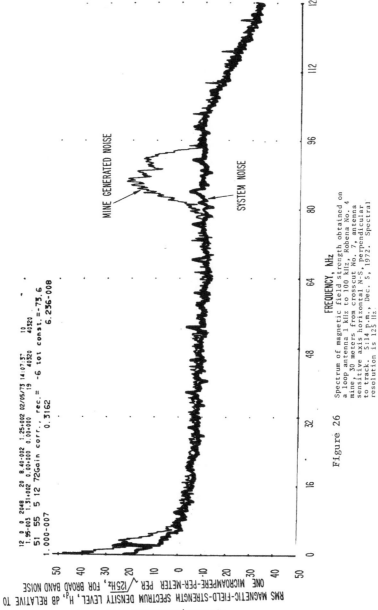

Figure 26

Spectrum of magnetic field strength obtained on
a loop antenna 1 kHz to 100 kHz, Robena No. 4
mine, 30 meters from crosscut No. 7, antenna
sensitive axis horizontal N-S, perpendicular
to track. 5:14 p.m., Dec. 5, 1972. Spectral
resolution is 125 Hz.

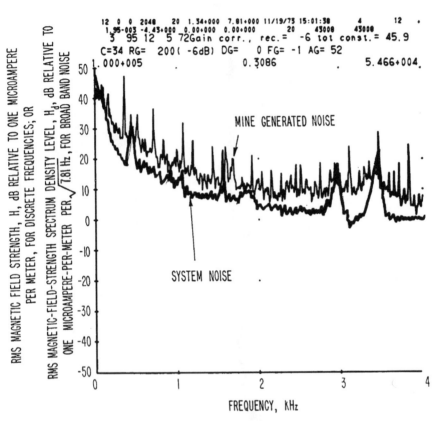

Figure 27   Spectrum of magnetic field strength obtained on
             a loop antenna 100 Hz to 3 kHz, Robena No. 4
             mine, underground, crosscut No. 7, 31.5 m from
             rail, antenna sensitive axis vertical, 5:14 p.m.,
             Dec. 5, 1972.   Spectral resolution is 7.81 Hz.

54

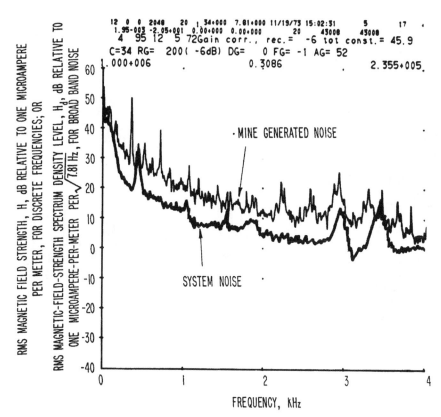

Figure 28   Spectrum of magnetic field strength obtained on
a loop antenna 100 Hz to 3 kHz, Robena No. 4
mine, underground, crosscut No. 7, 31.5 m from
rail, antenna sensitive axis horizontal E-W,
5:14 p.m., Dec. 5, 1972. Spectral resolution
is 7.81 Hz.

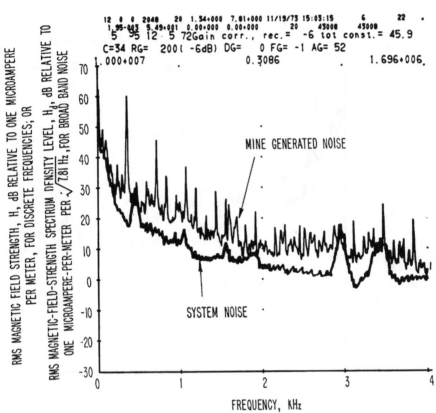

Figure 29   Spectrum of magnetic field strength obtained on
a loop antenna 100 Hz to 3 kHz, Robena No. 4
mine, underground, crosscut No. 7, 31.5 m from
rail, antenna sensitive axis horizontal N-S,
5:14 p.m., Dec. 5, 1972. Spectral resolution
is 7.81 Hz.

e. Nip Point (Location 4)

The intersection of the main haulage and crosscut 9 contains the "nip point," associated "power sled," and "car-pull," and is a central point of activity. The shuttle cars use this area to load (transfer) coal into the train cars. As the cars are filled, they are moved forward under the control of an operator using a large hydraulic car-pull to move in empty cars for filling. Antenna location number 4 is one meter away from the electric motor that drives the hydraulic pump for the car-pull. The antenna is located between the motor and the track in an area of almost constant occupancy by the car-pull operator. The antenna sensitive axis is horizontal E-W. Figure 30 shows the spectrum obtained with the car-pull motor operating. The car-pull operates intermittently every few seconds, for a few seconds, while a shuttle car is unloading. The maximum spectrum value is at 78 dB μA/m at 1000 Hz, drops to 47 dB μA/m at 10 kHz, and is down to 25 dB μA/m at 30 kHz.

This spectrum contains no spectral peaks due to brush noise, which is unusual for a dc motor. The sound produced in the audio monitor while recording contained no whine. Whine usually is associated with brush-produced spectral peaks. No explanation is apparent for this absence of peaks. Note that the field strength measured here at antenna location 4 is the highest field produced by a single machine. However, higher field strengths are measured at multiples of power line frequencies in cuts containing primary mine dc power cables.

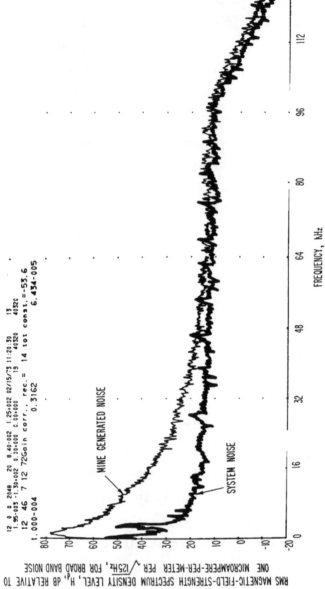

MINE GENERATED NOISE

SYSTEM NOISE

12 0 0 2048 20 8.40-002 1.25-002 02/15/73 11:20:30   13
1.95-005 -1.30-002 0.20-000 0.00-000   19   40320   40320
12 46   7 12 72Gain corr., rec. = 14 tot const.=-53.6
          1.000-004   0.3162   6.454-005

RMS MAGNETIC FIELD STRENGTH, h, dB RELATIVE TO ONE MICROAMPERE
PER METER, FOR DISCRETE FREQUENCIES; OR

RMS MAGNETIC-FIELD-STRENGTH SPECTRUM DENSITY LEVEL, H_d, dB RELATIVE TO
ONE MICROAMPERE-PER-METER PER $\sqrt{125Hz}$, FOR BROAD BAND NOISE

FREQUENCY, kHz

Figure 30   Spectrum of magnetic field strength obtained on
a loop antenna 1 kHz to 100 kHz, Robena No. 4
mine, underground, crosscut No. 9, antenna sensitive axis horizontal E-W, 1 m from car pull
electric motor, car pull operating. 12:52 p.m.,
Dec. 7, 1972. Spectral resolution is 125 Hz.

Other spectra taken at antenna location #4 showed the 88-kHz mine phone signal strength as 42 dB μA/m.

f. Air-split

An area considered important for a communication sub-station in the event of an emergency is an "air-split," that is, an area of the mine where two streams of fresh air converge or diverge. One air-split (not shown on report mine maps) in Robena No. 4 mine is at the intersection of 3-main with 10-right about 1 mile from the working section previously discussed. Figure 31 shows the spectrum taken at the air-split with the antenna (sensitive axis vertical) about 1 meter from the nearest rail. Figure 32 shows the expanded spectrum. In the region between 1080 and 2880 Hz the 3rd through 8th harmonics of 360 Hz are apparent. The levels of the harmonics are 80 dB μA/m for the 3rd harmonic, 76 dB for the 4th, 73 dB for the 5th, 68 dB for the 6th, 60 dB for the 7th, and 49 dB μA/m for the 8th harmonic. The logarithmic average strength of harmonics 3 through 8 is 67.8 dB μA/m. This compares with 26.0 dB μA/m at the face and 59.4 dB μA/m (for harmonics 3 through 8) in the section haulage way. Apparently larger mine dc supply currents are flowing through cables near this intersection with correspondingly higher (8.4 dB) logarithmic average powerline harmonic noise. Above the 8th harmonic (see figure 31), mine noise continues to decline, having a value

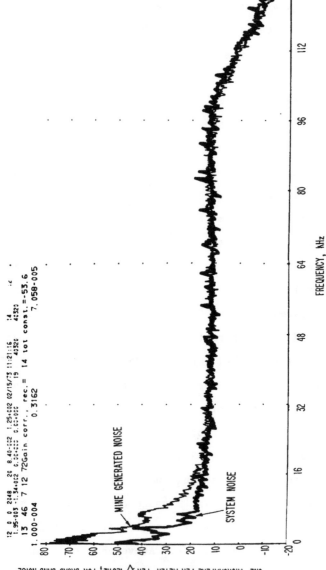

Figure 31 — Spectrum of magnetic field strength obtained on a loop antenna 1 kHz to 100 kHz, Robena No. 4 mine, underground, air split, antenna sensitive axis vertical, 1:11 p.m., Dec. 7, 1972. Spectral resolution is 125 Hz.

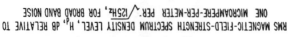

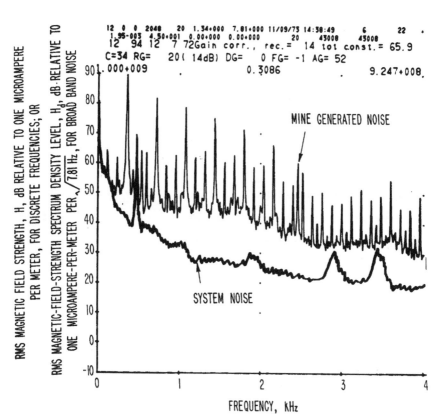

RMS MAGNETIC FIELD STRENGTH, H, dB RELATIVE TO ONE MICROAMPERE PER METER, FOR DISCRETE FREQUENCIES; OR

RMS MAGNETIC-FIELD-STRENGTH SPECTRUM DENSITY LEVEL, H_d, dB·RELATIVE TO ONE MICROAMPERE-PER-METER PER $\sqrt{7.81\,Hz}$, FOR BROAD BAND NOISE

MINE GENERATED NOISE

SYSTEM NOISE

FREQUENCY, kHz

Figure 32   Spectrum of magnetic field strength obtained on
             a loop antenna 100 Hz to 3 kHz, Robena No. 4
             mine, underground, antenna sensitive axis ver-
             tical, 1:11 p.m., Dec. 7, 1972. Air Split.
             Spectral resolution is 7.81 Hz.

61

41 dB μA/m at the 18th harmonic (648 kHz), 31 dB μA/m at the 25th harmonic (9 kHz), and going into the system noise of about 19 dB μA/m at about 16 kHz.

g.   Bailey Shaft (Location 6)

An entirely different area of the Robena No. 4 mine near Bailey Shaft, about 10-km away, was measured.   Figure 33 shows the features near Bailey Shaft.   The overburden in this area is 184 m (605 ft.).   Bailey shaft is an open shaft carrying fresh air for ventilation, a water pipe, and two heavy mine dc power cables.   The bottom of the shaft is several hundred meters from the nearest rail haulage.   Thirty meters east and 10 meters south of the bottom of Bailey shaft is a pump room containing a dc motor-driven pump for pumping water out of Robena No. 4.   Figure 34 shows the vertical sensitive axis spectrum taken for an antenna location (number 6) thirty meters east of the bottom of Bailey shaft.   Figure 35 shows the expanded spectrum.   The logarithmic average of the amplitudes of harmonics 3 through 8 is 66.1 dB μA/m.   The harmonic amplitudes remain fairly constant (between 58 and 70 dB μA/m) out to the 12th harmonic (4.32 kHz).   Beginning with the 13th harmonic, a dramatic and steady decline in harmonic amplitude occurs.   The 18th harmonic (6.48 kHz) is down to 37 dB μA/m.   The last clearly recognizable harmonic (before receiver noise becomes predominant) is the 25th (9 kHz) with an ampli-

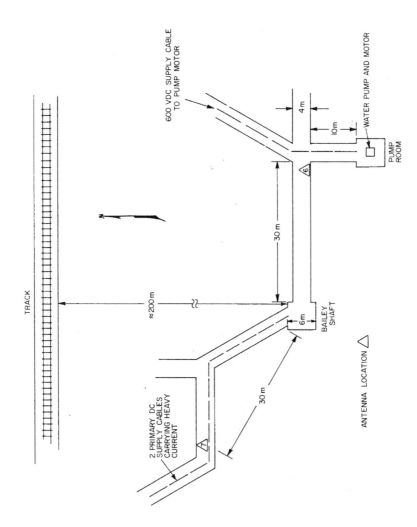

Figure 33 Map of antenna locations underground near Bailey shaft.

63

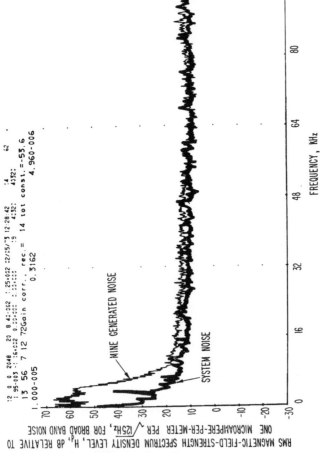

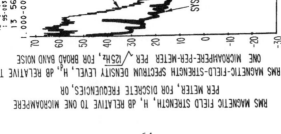

Figure 34

Spectrum of magnetic field strength obtained on
a loop antenna 1 kHz to 100 kHz, Robena No. 4
mine, underground, 30 meters east of bottom of
Bailey Shaft, 2:45 p.m., Dec. 7, 1972. Antenna
sensitive axis vertical. Spectral resolution
is 125 Hz.

64

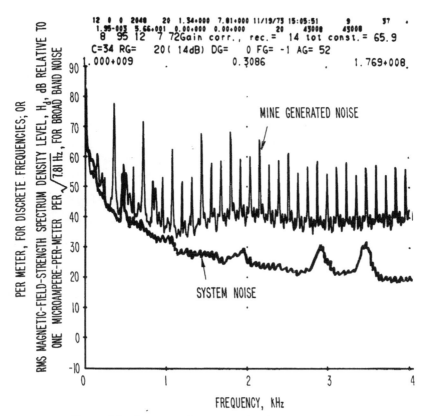

Figure 35   Spectrum of magnetic field strength obtained on
a loop antenna 100 Hz to 3 kHz, Robena No. 4
mine, underground, antenna sensitive axis vertical, 2:45 p.m., Dec. 7, 1972.   30 m east of the
bottom of Bailey shaft.   Spectral resolution is
7.81 Hz.

65

tude of 22 dB µA/m. Note that the logarithmic average amplitude is about the same as that found at the air-split (logarithmic average of 3 though 8 is 67.8 dB µA/m). The noise
measurements of primary interest are the ones taken at times
of maximum noise, since potential communications systems must
operate through this noise. To illustrate the variability of
the noise spectrum with time, figure 36 shows the noise spectrum taken a few seconds later. Figure 37 shows the expanded
spectrum. The logarithmic average of the amplitudes of
harmonics 3 through 8 is 55.8 dB µA/m, or lower by 10.3 dB.
Figure 38 shows the vertical sensitive axis noise spectrum
taken 27 minutes later at the same location. Figure 39 shows
the expanded spectrum. The logarithmic average amplitude of
harmonics 3 through 8 is about 45 dB µA/m, or lower by 21 dB.
Figure 38 also shows the amplitude of the 88-kHz trolley
phone as being 46 dB µA/m. The signal probably is propagated
along the 600-volt dc line supplying power to the water pump
motor about 10 meters away.

Because of the remoteness of Bailey shaft, recording
equipment sufficient for only one channel of data was carried
to the location. To measure the three orthogonal components
of noise at Bailey, the antenna had to be placed serially in
three different orientations. A set of three measurements
taken three minutes apart produced logarithmic-average noise
amplitudes (of harmonics 3 through 8) of 47.2, 35.0, and 49.2

66

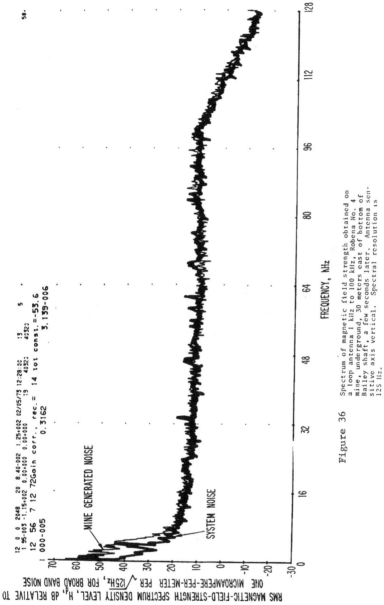

Figure 36   Spectrum of magnetic field strength obtained on
a loop antenna 1 kHz to 100 kHz, Robena No. 4
mine, underground, 30 meters east of bottom of
Bailey shaft, a few seconds later. Antenna sensitive axis vertical. Spectral resolution is
125 Hz.

67

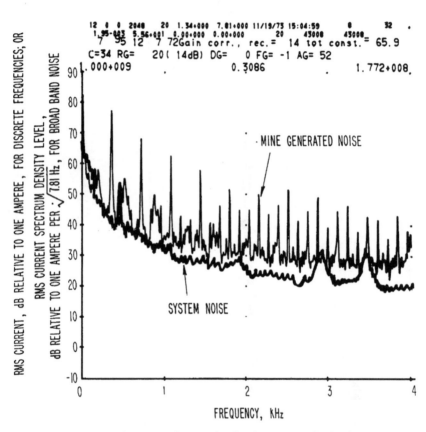

Figure 37   Spectrum of magnetic field strength obtained on
a loop antenna 100 Hz to 3 kHz, Robena No. 4
mine, underground, antenna sensitive axis vertical, Dec. 7, 1972. 30 m east of the bottom of
Bailey shaft. A few seconds later. Spectral
resolution is 7.81 Hz.

68

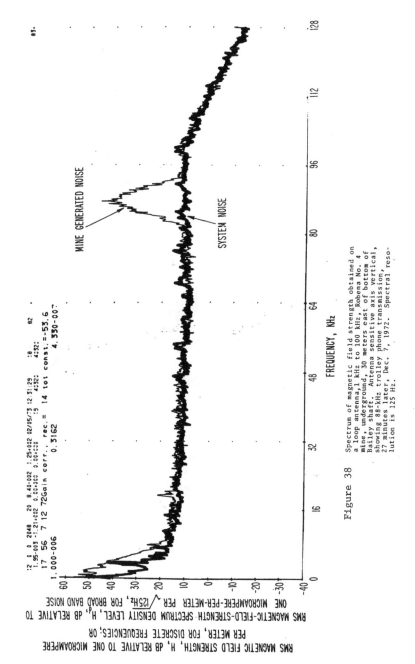

Figure 38   Spectrum of magnetic field strength obtained on
a loop antenna, 1 kHz to 100 kHz, Robena No. 4
mine underground, 30 meters east of bottom of
Bailey shaft. Antenna sensitive axis vertical,
showing 88-kHz trolley phone transmission,
27 minutes later, Dec. 7, 1972. Spectral resolution is 125 Hz.

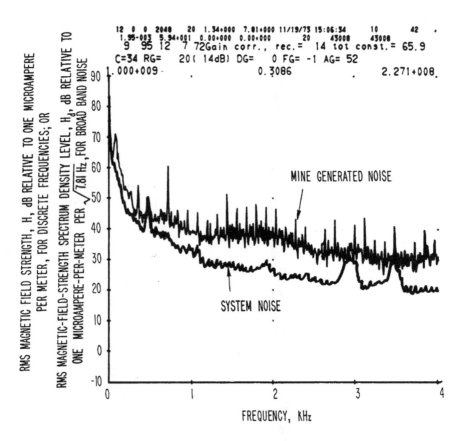

Figure 39   Spectrum of magnetic field strength obtained on
            a loop antenna 100 Hz to 3 kHz, Robena No. 4
            mine, underground, antenna sensitive axis verti-
            cal, Dec. 7, 1972. 30 m east of the bottom of
            Bailey shaft, 27 minutes later. Spectral
            resolution is 7.81 Hz.

70

dB µA/m for horizontal N-S, horizontal E-W, and vertical antenna sensitive axes, respectively. Typically, the vertical sensitive axis is strongest. The spectra are similar to those shown previously and are not included here.

h. Bailey Shaft (Location 7)

A final magnetic field strength measurement was made at antenna location 7, 30 meters in another direction from the bottom of Bailey shaft in a cut carrying the primary 600 volt dc supply cables for a large portion of the mine. The antenna sensitive axis is vertical, and the antenna is placed about 2 meters from the cables carrying heavy currents. The primary purpose of this particular antenna placement is to obtain a coherence of variations in noise simultaneously on the surface and underground. The results will be discussed in section IV, A. Absolute field strengths are obtained from this measurement by the usual calibration and correction techniques. The spectral analysis bandwidth (62.5 Hz) and the frequency range (300 Hz to 40 kHz) are different than past data analyses. Figure 40 shows the field strength measured. The field strengths are unusually high. The logarithmic average of harmonics 3 through 8 is 98.2 dB µA/m. Subtracting 26.0 dB µA/m (the logarithmic-average harmonic strength measured at the face during quiet time) gives a 72-dB range in magnetic field strengths for these power line harmonics in Robena No. 4 mine.

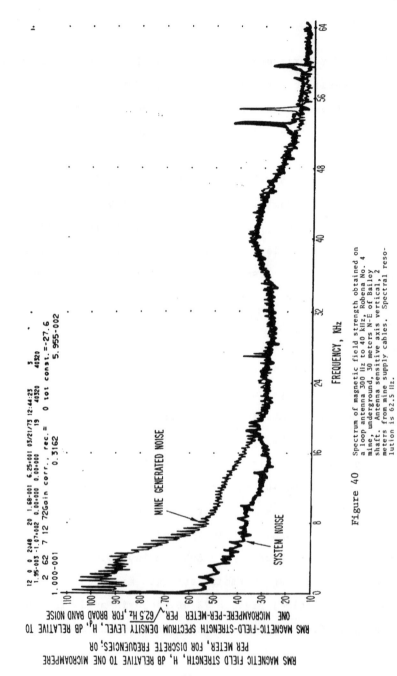

Figure 40   Spectrum of magnetic field strength obtained on
a loop antenna 300 Hz to 40 kHz, Robena No. 4
mine, underground, 30 meters N-E of Bailey
shaft. Antenna sensitive axis vertical, 2
meters from mine supply cables. Spectral resolution is 62.5 Hz.

This large range corresponds to an amplitude ratio of 5000 or
a power ratio of 25,000,000.   In summary, a coal mine can be very
quiet or very noisy (electrically), depending on where measurements are taken relative to operating electrical cables and
equipment.

i.   Summary Plot of Power Line Harmonics

Figure 41 is a summary of magnetic field strength at
power-line harmonic frequencies observed within Robena No. 4.
Plotted are the logarithmic averages of harmonics 3 through 8
of 360 Hz (i.e., 1080 Hz through 2880 Hz).   Impulsive noise
is not shown.   For comparison, the two diamond-shaped points
show equipment-generated noise.

III.   AMPLITUDE PROBABILITY DISTRIBUTION MEASUREMENTS

A.   Introduction

The amplitude probability distribution (APD) of the received noise signal magnitude is one of the more useful
statistical descriptions of the noise process for the design
and evaluation of a telecommunications system operating in a
noisy environment [7, 8, 9].

By plotting the cumulative APD on Rayleigh graph paper,
one can show clearly the fraction of time that the noise exceeds various levels.   We use Rayleigh graph paper with scales

73

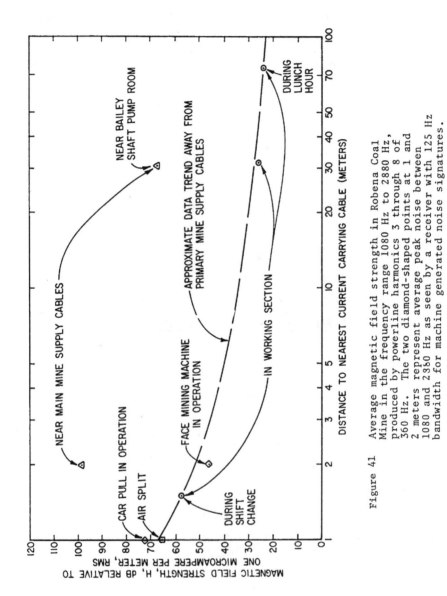

Figure 41 Average magnetic field strength in Robena Coal
Mine in the frequency range 1080 Hz to 2880 Hz,
produced by powerline harmonics 3 through 8 of
360 Hz. The two diamond-shaped points at 1 and
2 meters represent average peak noise between
1080 and 2380 Hz as seen by a receiver with 125 Hz
bandwidth for machine generated noise signatures.

74

chosen so that Gaussian noise (e.g., thermal noise) plots as a straight line with slope of -1/2. Noise with rapid large changes in amplitude (e.g., impulsive noise) then has a much steeper slope, typically -4 or -5, depending on the receiver bandwidth.

Section III of this report describes the APD measurement methods and results. Part B.1 describes the measurement instrumentation of an underground recording system, a data transcribing system, and a data processing system. Part B.2 presents measurement techniques used for APD, rms, and average measurements of noise in a coal mine. Part B.3 describes the calibration procedure and estimate of accuracy in our measurements. Part C includes many APD's taken in a coal mine. An analysis of these results also is included.

B. Noise Measurement Techniques

1. Measurement Instrumentation

Section III, B of this report describes the system used to measure the amplitude probability distribution statistics of electromagnetic noise in coal mines. The system is an extension of one designed by Matheson [10]. See figures 42a, b, and c.

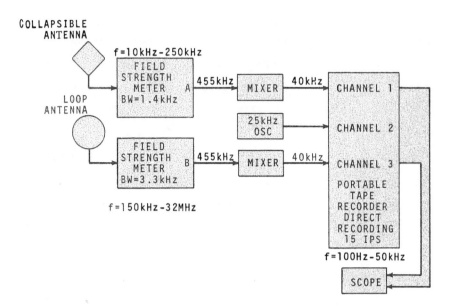

COLLAPSIBLE
ANTENNA

f=10kHz-250kHz

FIELD STRENGTH METER A
BW=1.4kHz

455kHz

MIXER

40kHz

CHANNEL 1

LOOP
ANTENNA

25kHz OSC

CHANNEL 2

FIELD STRENGTH METER B
BW=3.3kHz

455kHz

MIXER

40kHz

CHANNEL 3

f=150kHz-32MHz

PORTABLE TAPE RECORDER DIRECT RECORDING 15 IPS

f=100Hz-50kHz

SCOPE

Figure 42a   System for field recording.

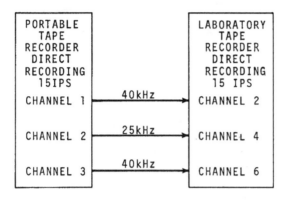

Figure 42b   System for transcribing.

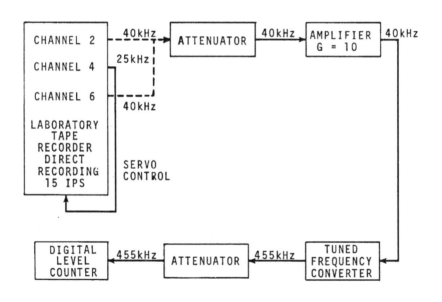

Figure 42c   System for ADP processing.

77

a. Underground Recording System

The principal parameter measured is magnetic field
strength. Electrostatically shielded loop antennas are used
to intercept the magnetic field and to discriminate substan-
tially against any electric-field component. For the frequency range between 10 kHz and 250 kHz, the loop antenna is
a collapsible, single-turn diamond configuration of area of
about 0.7 square meters. This loop has an inductance of
about a microhenry, and at low frequencies represents a very
low impedance source compared to the 50-ohm input impedance
of the field strength meter. Therefore, a balun with step-up
transformer is used to match the low impedance antenna to the
50-ohm input. For the frequency range between 150 kHz and
32 MHz, a single-turn, 38-cm diameter, circular loop antenna
is used with a balun. This loop antenna is also electrically
shielded and has an inductance of about a microhenry. The
magnitude of the impedance of the loop antenna varies from
one ohm at 150 kHz to 200 ohms at 32 MHz. A switch on each
balun allows use of several impedance-matching networks (four
for the low frequency case and eight in the high frequency
case), which consist of transformers and coupling capacitors
to give the desired match over the required frequency range.
The outputs of the baluns are fed into commercially available,
battery-powered, electromagnetic interference and field strength
meters (hereafter referred to as EIFS meters).

78

It has long been recognized that a mean square measure is a very useful statistical measure. But although many EIFS meters have detector functions such as peak, quasi-peak, and average voltage, very few of them have the very important function of rms voltage. The EIFS meters used for our electromagnetic noise measurements are modified to measure rms voltage simultaneously with average voltage [10]. The characteristics of these modified EIFS meters used for our noise measurements are listed in Table 1.

The particular system that we used actually measures rms voltage, V_{rms}, and average voltage, V_{avg}. The system functions by automatically adjusting the receiver gain to keep a constant rms voltage at the output of the integrator following the squared-voltage detector. The receiver has a logarithmic gain control characteristic. One can examine the automatic gain control (AGC) voltage and obtain directly the input rms voltage in dB.

In order to measure the APD, the AGC circuit is disabled. Since the gain of the receiver is now constant, the magnitude of the IF output is directly related to the bandlimited input noise signal magnitude. To check the linearity and to establish the dynamic range of the EIFS meter, a CW signal is applied to the input of the EIFS meter. The relation between the input and the IF output of the EIFS meter A is shown in figure 43. The dynamic range of the EIFS meter used in our noise measurements at 1.4-kHz bandwidth is found to be 65 dB.

79

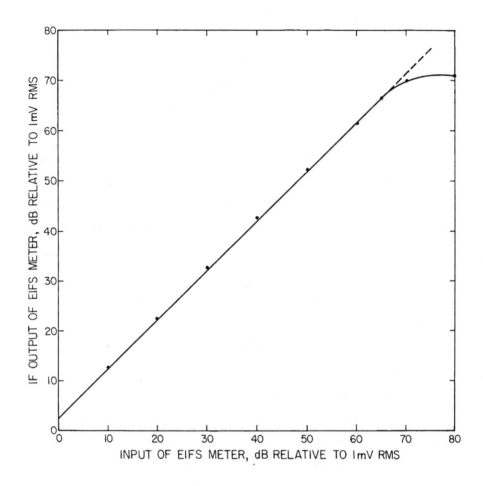

Figure 43   Linearity of Electromagnetic Interference and Field
Strength (EIFS) Meter.

80

Table 1. Characteristics of Electromagnetic Interference
and Field Strength (EIFS) Meters
(Manufacturer's Specifications)

EIFS Meter	A	B
Frequency Coverage	10 kHz - 250 kHz	150 kHz - 32 MHz
Sensitivity	0.01 µV	0.1 µV
cw 3-dB Signal Bandwidth*	1.4 kHz	3.3 kHz
Dynamic Range	67 dB	60 dB
Spurious Response Rejection	>50 dB	>60 dB
Input Impedance	50 ohms with an input VSWR less than 1.2:1	50 ohms with an input VSWR less than 41.2:1

*These values of the cw 3-dB signal bandwidth are based on
NBS measurements.

Since the bandwidth of the portable tape recorder is limited to 50 kHz, the IF output from the EIFS meter is converted down from 455 kHz to 40 kHz using mixers. The circuit diagram for the mixers used in our noise measurements is shown in figure 44. A commercially available, battery-powered, portable, analog magnetic tape recorder is used for recording. The tape speed chosen on record and on playback is 15 inches per second (ips). At this speed the portable tape recorder frequency response range is 100 Hz to 50 kHz at the ± 2 dB points in the direct recording mode as shown in figure 45. The input voltage range is adjusted to record the signal level between 10 millivolts and 1 volt rms. The tape recorder gain is adjusted for 0 dB. The characteristics of this portable tape recorder are listed in Table 2.

An external set of sealed, lead-acid batteries in an explosion-proof enclosure is used to drive the portable tape recorder. The current is limited by a solid-state, current-limiting circuit in series with a fuse. The power requirement is approximately 13 watts at a nominal 17.5 volts. This battery system allows about eight hours of recording.

b. Data Transcribing System

The cumulative peak-to-peak flutter of the portable tape recorder is about 0.8 percent, whereas that of the laboratory tape recorder is about 0.4 percent. The time displacement error is perhaps more important, being microseconds for the

82

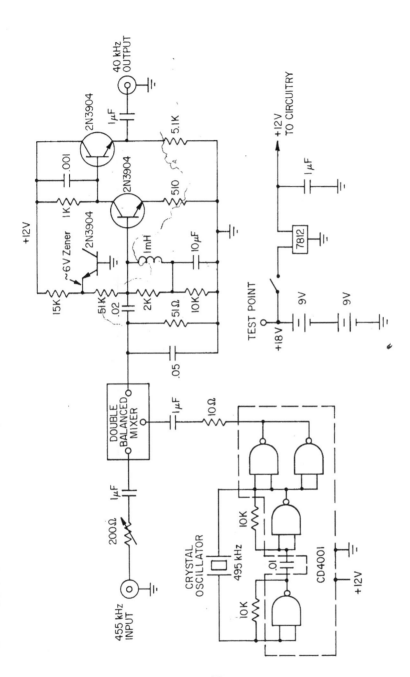

Figure 44 Mixer circuit diagram.

83

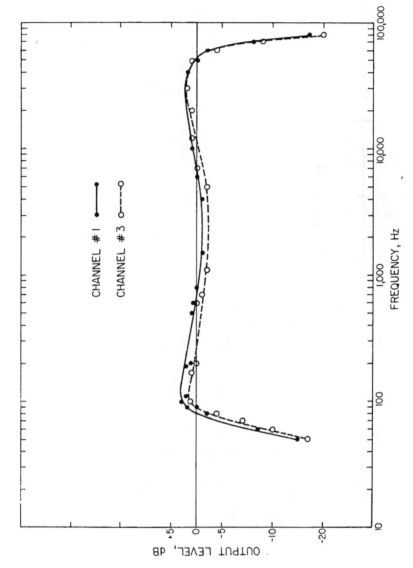

Figure 45   Frequency response of portable analog magnetic tape recorder.

Table 2. Characteristics of Portable Magnetic Tape Recorder
(Direct Record/Reproduce)
(Manufacturer's Specifications)

Tape Speed	15 inches per second (ips)
Flutter (cumulative peak-to-peak flutter)	0.8 percent
rms Signal to rms Noise	35 dB
Crosstalk	35 dB below nominal signal level
Harmonic Distortion	1.5 percent total harmonic distortion
Input Voltage Range	0.01 to 1 V rms
Input Impedance	20 kΩ
Output Voltage	1 V rms into 1-kΩ load
Output Impedance	100 Ω
Frequency Response at 15 ips	100 Hz to 50 kHz at ± 2-dB points

laboratory tape recorder and milliseconds for the portable tape recorder. Therefore later, at our laboratory, the tapes are transcribed through a laboratory tape recorder whose servo system can take out the flutter and wow introduced by the portable tape recorder. The characteristics of this laboratory tape recorder are listed in Table 3. To give a reference time base, a stable 25-kHz signal is recorded on a separate track at the time the mine recordings are made. At playback time (after transcription) this signal is used to control the servo of the laboratory tape recorder.

c. Data Processing System

The data processing system consists principally of the laboratory analog magnetic tape recorder as a playback unit, an amplifier, a tuned frequency converter, and a digital level counter. The amplifier is used primarily for impedance conversion between the output impedance of the laboratory tape recorder and the input impedance of the tuned frequency meter. The 40-kHz output of the laboratory tape recorder is converted up to 455 kHz by the tuned frequency converter in order to match the response band of the digital level counter.

The digital level counter provides a direct digital display of the percentage of the time each of 15 levels, 6 dB apart, are exceeded. This instrument has five identical 18-dB amplifiers in series. Each amplifier has a detector on its

86

Table 3.

Characteristics of Laboratory Analog Magnetic Tape Recorder
(Direct Record/Reproduce)
(Manufacturer's Specifications)

Tape Speed	15 inches per second (ips)
Flutter (cumulative peak-to-peak flutter)	0.4 percent
rms Signal to rms Noise	37 dB
Harmonic Distortion	1 percent
Input Voltage Range	0.3 to 3.0 V rms
Input Impedance	100 kΩ
Output Voltage	1.0 V rms into 10-kΩ load
Output Impedance	< 100 Ω
Frequency Response at 15 ips	100 Hz to 75 kHz at ± 3-dB points

output. Each detector is linear over an 18-dB range, and its output is discriminated to be in one of three levels, six dB apart. Each level, whenever exceeded, drives a Schmitt trigger which gates a clock on; the gated clock pulses are counted in parallel in each of 15 (three times five) independent counters. This gives a cumulative distribution. The lower-level channels generally read nearly 100 percent. A 16th counter reads the corresponding total clock pulses. Readout is on a 7-digit display which has a maximum of 9×10^6 counts available for each channel. The clock rate can be adjusted from 1 kHz to 1 MHz in 1-2-5 steps.

The cw, 3-dB signal bandwidth of the whole system, including the recording, transcribing, and data processing systems, is primarily determined by the data processing system. The predetection bandwidth of the APD measurements for the frequency range between 1.0 kHz and 250 kHz using EIFS meter A is 10 kHz, whereas that for the frequency range between 250 kHz and 32 MHz using EIFS meter B is 1.2 kHz. These predetection bandwidths are indicated in each APD figure. The dynamic range of the whole system is primarily limited by the magnetic tape recorder to about 45 dB.

The system used for recording, transcribing, and data processing is shown in figures 42a, b, c. Figure 46 shows a collapsible loop antenna used in a mine for the frequency range between 10 kHz and 250 kHz. The recording system which includes

Figure 46   Collapsible loop antenna as used in Robena mine for frequencies between 10 kHz and 250 kHz.

two EIFS meters, two mixers, a portable analog magnetic tape recorder, an external set of sealed, lead-acid batteries, a portable oscilloscope, etc., used in a mine is shown in figure 47. Figure 48 shows the data processing system which consists of a laboratory analog magnetic tape recorder, an amplifier, a tuned frequency converter and a digital level counter.

2. Measurement Technique

The measurement technique is to record on magnetic tape a time-varying analog signal whose amplitude (envelope) varies proportionally to magnetic field strength, as seen through a specific receiver bandwidth. These signals are recorded for about 20 minutes at each frequency and for each of three antenna orientations. The tape is transcribed later through a laboratory tape recorder whose servo system can take out the flutter and wow introduced by the portable tape recorder. The data processing system consists principally of the laboratory tape recorder for a playback unit and a digital level counter which provides a direct digital display of the percentage of the time each of 15 levels, 6-dB apart, is exceeded.

The voltage levels applied to the recorder must be adjusted (by controlling the gain of the field strength meters) so that the curved portion of the APD falls within the dynamic range of the magnetic tape recorder. To do this, we set the main function switch to "noise" position and record rms and average

90

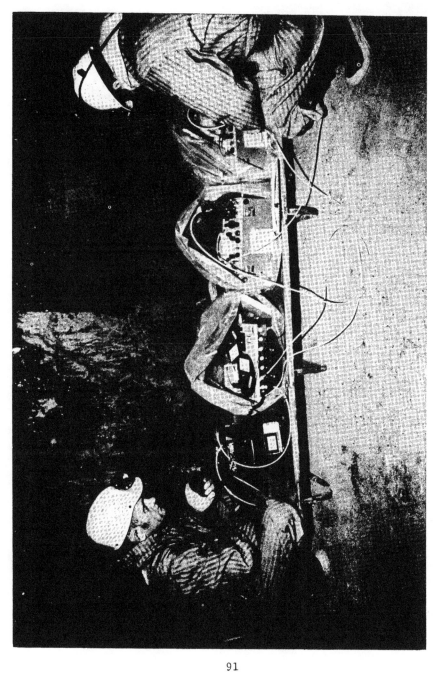

Figure 47   APD recording system in operation in Robena mine.

91

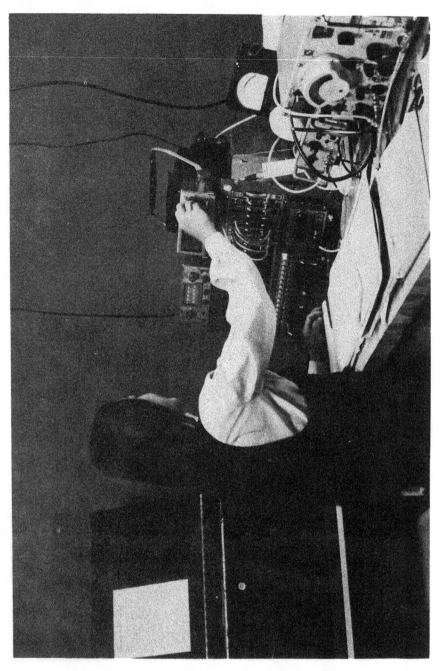

voltages. These two readings give true rms and average
values of noise in a coal mine on a fast-response basis. The
time constants for our measurements can be adjusted between
0.1 and 100 sec. depending on the circumstances. One can use
the shorter time constant for a Gaussian noise environment,
but longer time constants are required for an impulsive noise
environment. These "short term" rms and average values are
recorded and are used to set the gain of EIFS meters, but
these values are not reported here. We then set the main
function switch to peak position in order to disable the AGC.
By changing the gain of each EIFS meter, the IF output of each
EIFS meter is adjusted to about 100 mv peak to peak. A portable oscilloscope is used for monitoring. The necessity for
monitors is discussd earlier in this report.

3. Calibration

The calibration of the entire measurement system, including the loop antennas, field strength meters, mixers, magnetic
tape recorders, impedance transforming amplifiers, and the
digital level counter, is performed by immersing the receiving
loop antennas in a known field, generated at the NBS field
strength calibration site. Thus all levels of field strength
are given in absolute units. This technique is called the
standard field method. It is used to calibrate loop antennas
at NBS from 30 Hz to 30 MHz. The ratio (calibration factor)

of the known field strength to the output of the unknown
system is calculated.

This standard field, a cw, quasi-static near-zone magnetic
field, is produced by a single-turn, unshielded, balanced
transmitting loop of known radius carrying a known current.
The magnitude of the field at the receiving loop, produced by
a single-turn circular transmitting loop, is given by the
following equation [11, 12, 13]:

$$H = \frac{r_1^2 \, I}{2 \, (d^2 + r_1^2 + r_2^2)^{3/2}} \sqrt{1 + \left(\frac{2\pi d}{\lambda}\right)^2} \, , \qquad (1)$$

where H is the magnetic field strength in rms amperes per
meter,

   r_1 is the radius of transmitting loop in meters,

   r_2 is the radius of receiving loop in meters (if the
      receiving loop is rectangular, use the radius of a
      circle having the same area),

   d is the axial spacing in meters between the two coaxial
      loops,

   I is the transmitting loop current in rms amperes, and

   λ is the free-space wavelength in meters.

The transmitting and receiving loops are positioned co-
axially with respect to each other at a spacing of 1 to 2 meters.
The spacing is determined by the desired magnitude of the calibrating field and the frequency. Equation (1) is valid for

determining the magnetic field strength only when r_1, r_2, and d are small compared to λ. The loop spacing should be at least four times the radius of the larger of r_1 and r_2 for equation (1) to be valid within one percent.

The calibration site should be in an area that is free of sizeable metallic objects that might influence or distort the calibrating field. Normally, if the calibrating area is cleared of metallic objects within two or three times the loop spacing, d, there will be no appreciable effect. The calibrating site should, therefore, be in an area relatively free of overhead power lines, steel-reinforced walls, or shielded enclosures. A non-metallic building with no overhead wiring makes a satisfactory calibration site.

The estimated limits of error for our APD noise measurements are ± 5 dB. Several sources of error that are critical to the overall accuracy of our measurements are listed below:

1. Use of a discrete digital level counter (levels are 6 dB apart) contributes ± 3-dB quantization error limit. This ± 3-dB quantization error would be improved to ± 1-dB error by an improved calibration method, and future APD noise data will be reported with improved accuracy.

2. The entire system, i.e., recording, data transcribing, and data processing, has a calibration uncertainty of ± 0.5 dB.

3. The estimated uncertainty involved in using the portable and the laboratory tape recorders for record and playback is ± 0.5 dB due to harmonic distortion, flutter, dropout, cross-talk, gain instability, etc.

4. The gain instability during measurements, gain changes between measurements and calibration, and the non-linearity of EIFS meters and mixers, all combined, contribute ± 0.5 dB uncertainty.

5. The gain instability and non-linearity of the digital level counter, the tuned frequency converter, the amplifier, and attenuators, all combined, contribute ± 0.5 dB uncertainty.

C. Noise Measurement Results: APD's

Many APD's of magnetic field noise were taken during actual operation of the coal mine on December 5th and 7th, 1972. The loop antennas were placed about 300 meters from the face area (location 5 in figure 5). Three orthogonal components of magnetic field were measured at eight frequencies ranging from 10 kHz to 32 MHz. These frequencies are 10 kHz, 30 kHz, 70 kHz, 130 kHz, 500 kHz, 2 MHz, 8 MHz, and 32 MHz. The length of time for each measurement was 23 minutes.

The measured data are presented in 32 APD's. These APD's are given as figures 49 through 80. The vertical axis gives magnetic field strength, H, while the horizontal axis gives

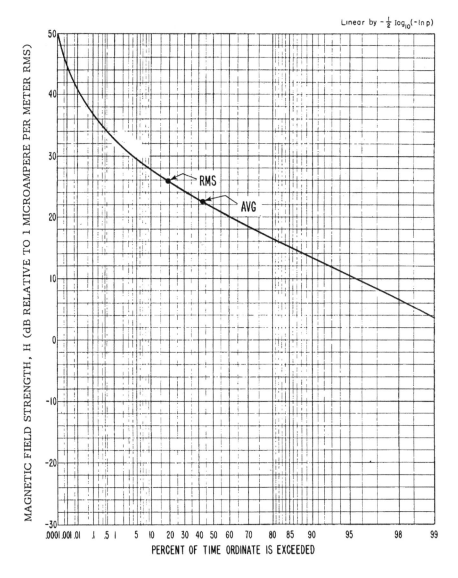

Figure 49   APD, 10 kHz, vertical component, 1.0 kHz predetection bandwidth, December 5, 1972, 11:25 a.m., Robena No. 4.

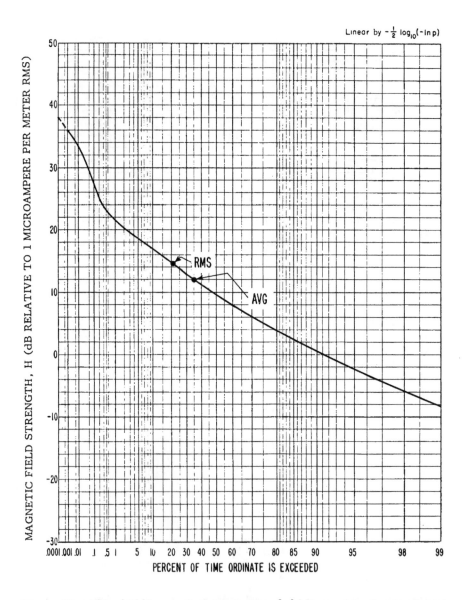

Figure 50   APD, 30 kHz, vertical component, 1.0 kHz predetection bandwidth,
December 5, 1972, 2:00 p.m., Robena No. 4.

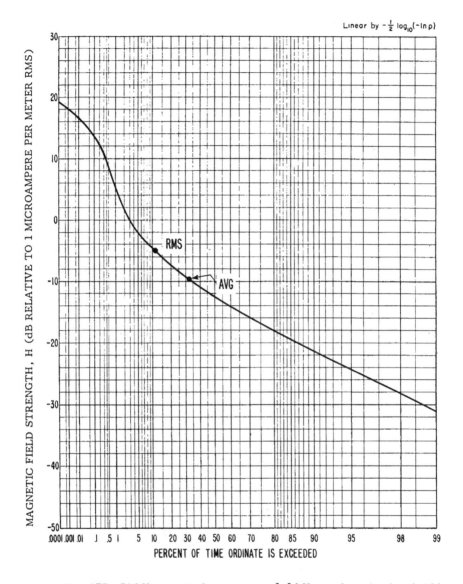

Figure 51   APD, 70 kHz, vertical component, 1.0 kHz predetection bandwidth,
December 5, 1972, 2:45 p.m., Robena No. 4.

99

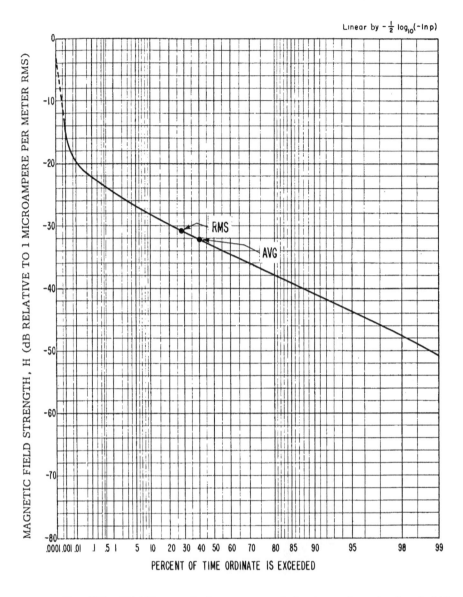

Figure 52   APD, 130 kHz, vertical component, 1.0 kHz predetection bandwidth, December 5, 1972, 5:45 p.m., Robena No. 4.

100

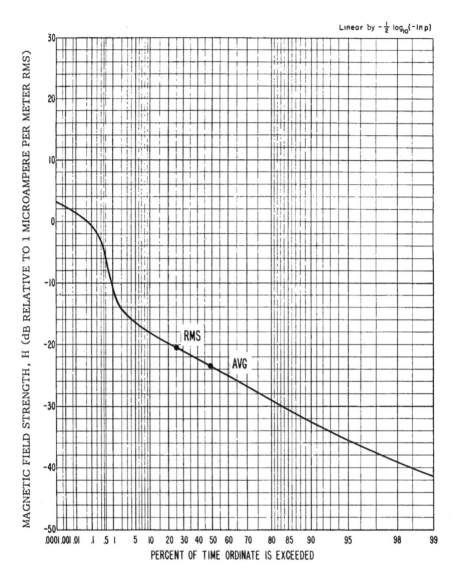

Figure 53   APD, 500 kHz, vertical component, 1.2 kHz predetection bandwidth, December 5, 1972, 12:45 p.m. Robena No. 4.

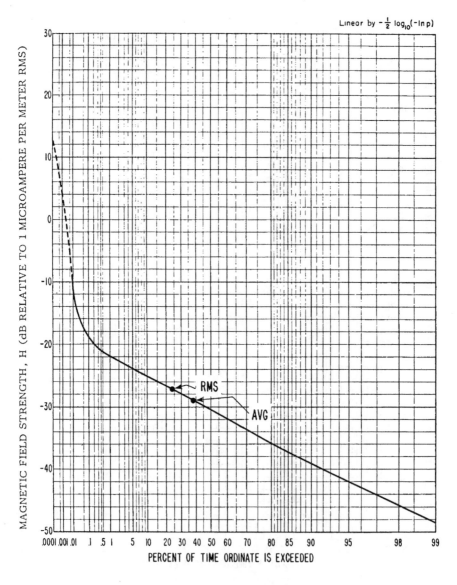

Figure 54   APD, 2 MHz, vertical component, 1.2 kHz predetection bandwidth,
December 5, 1972, 2:35 p.m., Robena No. 4.

102

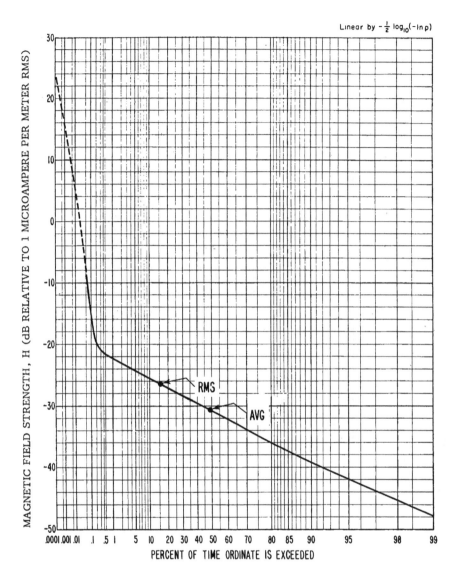

Figure 55 APD, 8 MHz, vertical component, 1.2 kHz predetection bandwidth, December 5, 1972, 4:45 p.m., Robena No. 4.

103

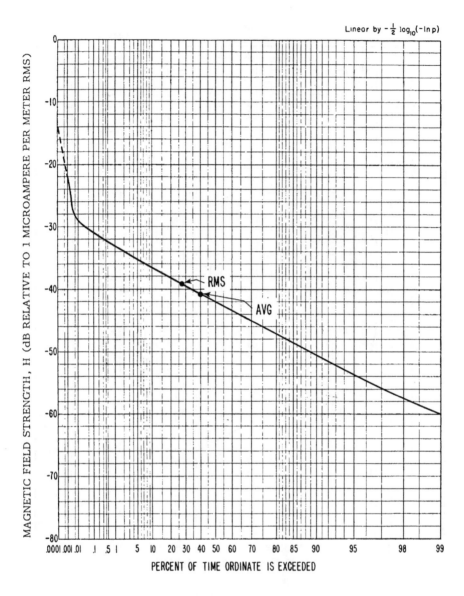

Figure 56   APD, 32 MHz, vertical component, 1.2 kHz predetection bandwidth, December 5, 1972, 6: 15 p.m., Robena No. 4.

104

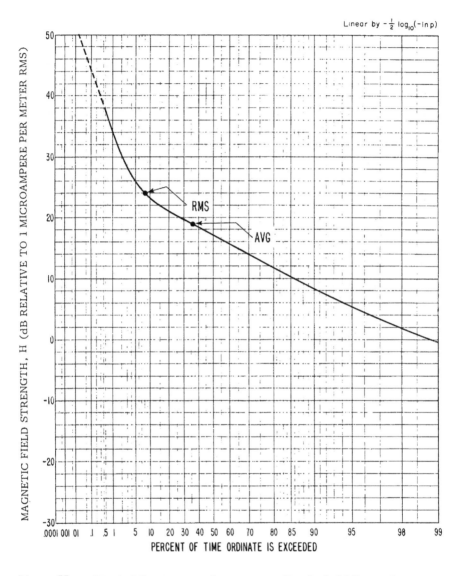

Figure 57   APD, 10 kHz, horizontal component (E-W), 1.0 kHz predetection bandwidth, December 5, 1972, 12: 45 p.m., Robena No. 4.

105

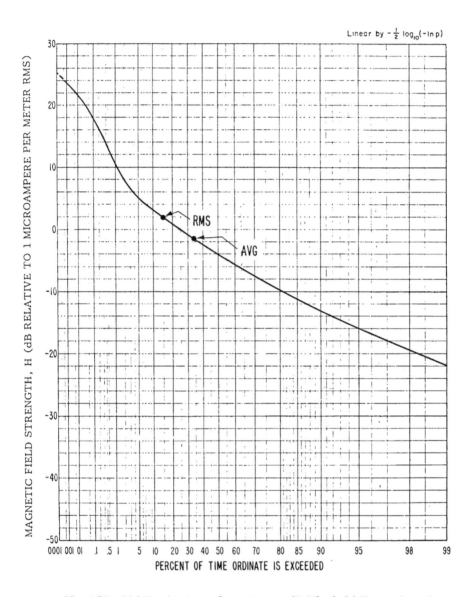

Figure 58   APD, 30 kHz, horizontal component (E-W), 1.0 kHz predetection
bandwidth, December 5, 1972, 2:30 p.m., Robena No. 4.

106

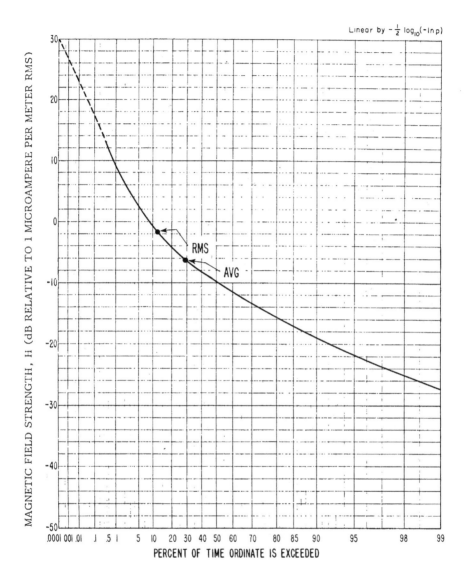

Figure 59   APD, 70 kHz, horizontal component (E-W), 1.0 kHz predetection
bandwidth, December 5, 1972, 4:45 p.m., Robena No. 4.

107

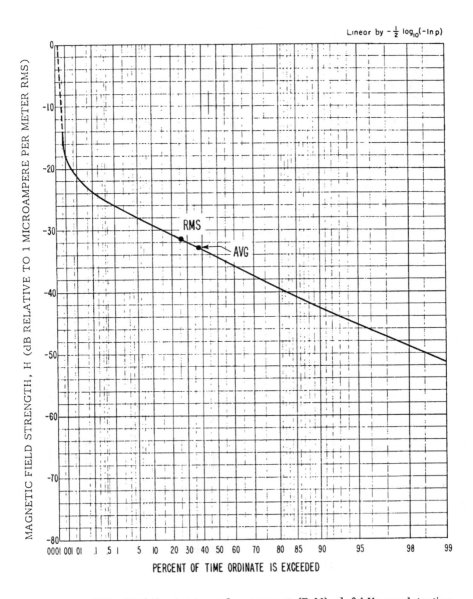

Figure 60   APD, 130 kHz, horizontal component (E-W), 1.0 kHz predetection
bandwidth, December 5, 1972, 6: 15 p.m., Robena No. 4.

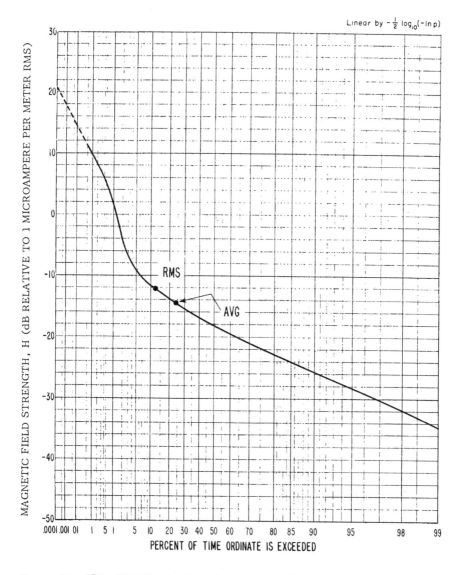

Figure 61   APD, 500 kHz, horizontal component (E-W), 1.2 kHz predetection
bandwidth, December 5, 1972, 11:25 a.m., Robena No. 4.

109

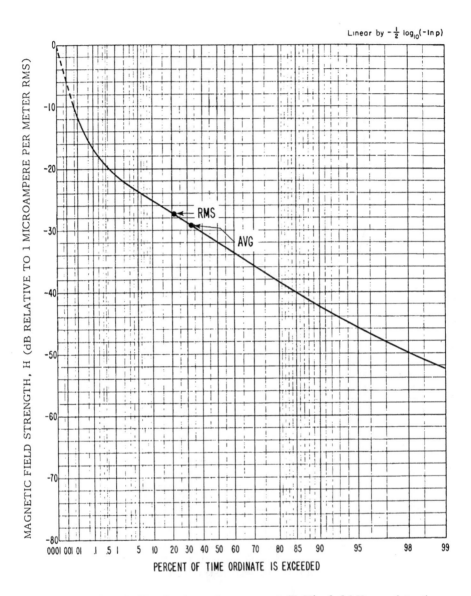

Figure 62   APD, 2 MHz, horizontal component (E-W), 1.2 kHz predetection
bandwidth, December 5, 1972, 2:00 p.m., Robena No. 4.

110

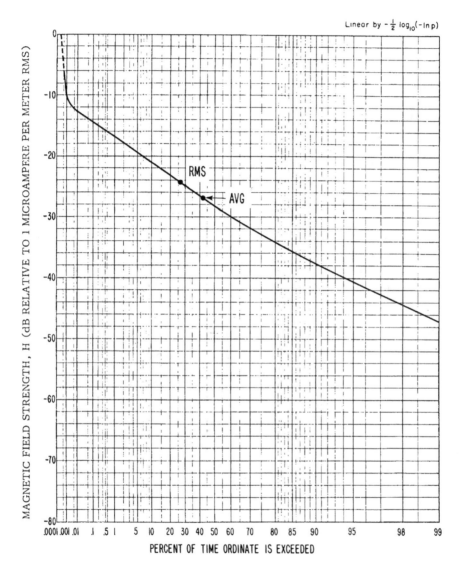

Figure 63   APD, 8 MHz, horizontal component (E-W), 1.2 kHz predetection
bandwidth, December 5, 1972, 5: 45 p.m., Robena No. 4.

111

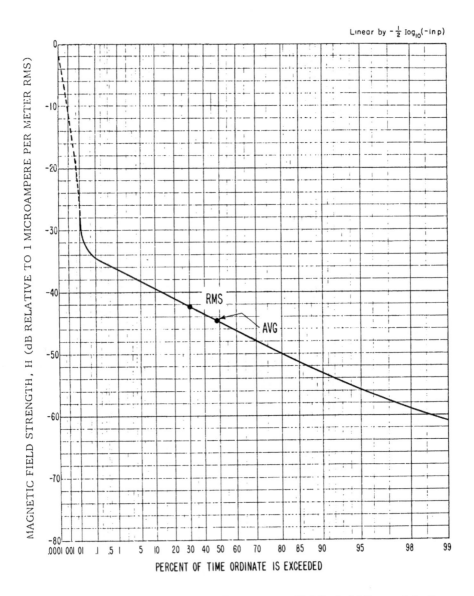

Figure 64   APD, 32 MHz, horizontal component (E-W), 1.2 kHz predetection bandwidth, December 5, 1972, 5:45 p.m., Robena No. 4.

112

B-74-1318

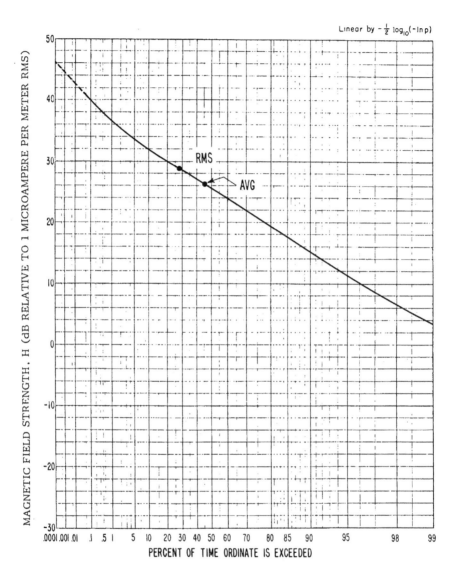

Figure 65   APD, 10 kHz, horizontal component (N-S), 1.0 kHz predetection
bandwidth, December 5, 1972, 1:20 p.m., Robena No. 4.

113

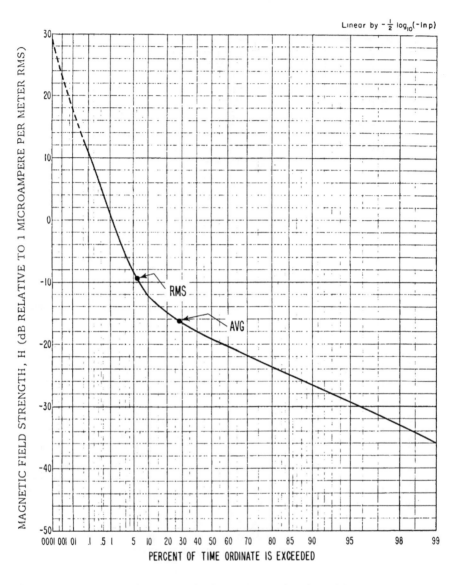

Figure 66   APD, 30 kHz, horizontal component (N-S), 1.0 kHz predetection
bandwidth, December 5, 1972, 3:05 p.m., Robena No. 4.

114

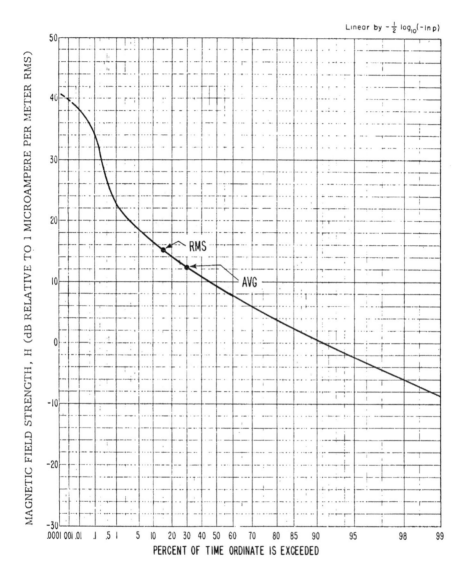

Figure 67   APD, 70 kHz, horizontal component (N-S), 1.0 kHz predetection bandwidth, December 5, 1972, 5:20 p.m., Robena No. 4.

115

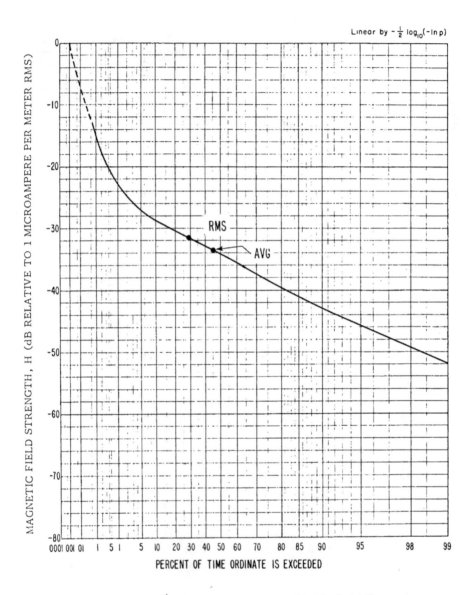

Figure 68   APD, 130 kHz, horizontal component (N-S), 1.0 kHz predetection
bandwidth, December 5, 1972, 6:50 p.m., Robena No. 4.

116

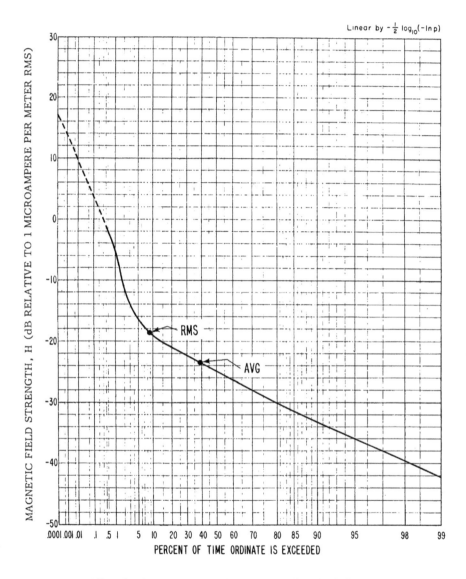

Figure 69   APD, 500 kHz, horizontal component (N-S), 1.2 kHz predetection
bandwidth, December 5, 1972, 1:30 p.m., Robena No. 4.

117

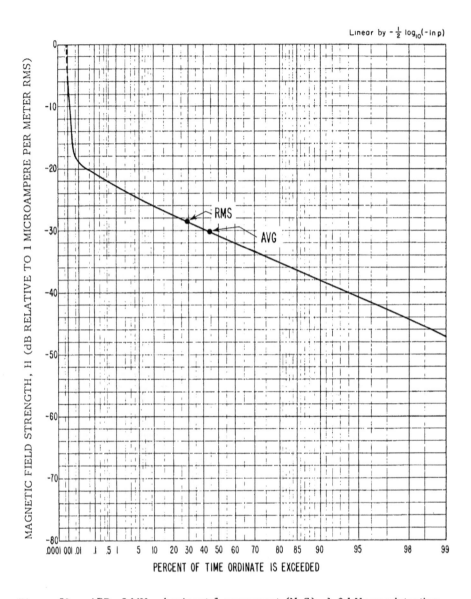

Figure 70   APD, 2 MHz, horizontal component (N-S), 1.2 kHz predetection
bandwidth, December 5, 1972, 3:05 p.m., Robena No. 4.

118

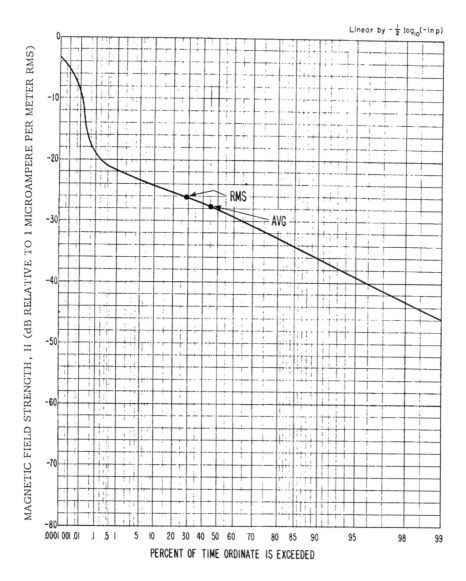

Figure 71   APD, 8 MHz, horizontal component (N-S), 1.2 kHz predetection
bandwidth, December 5, 1972, 5:20 p.m., Robena No. 4.

119

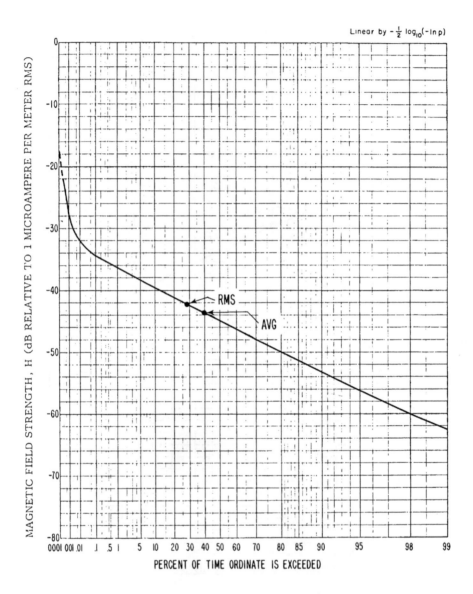

Figure 72   APD, 32 MHz, horizontal component (N-S), 1.2 kHz predetection
bandwidth, December 5, 1972, 6:50 p.m., Robena No. 4.

120

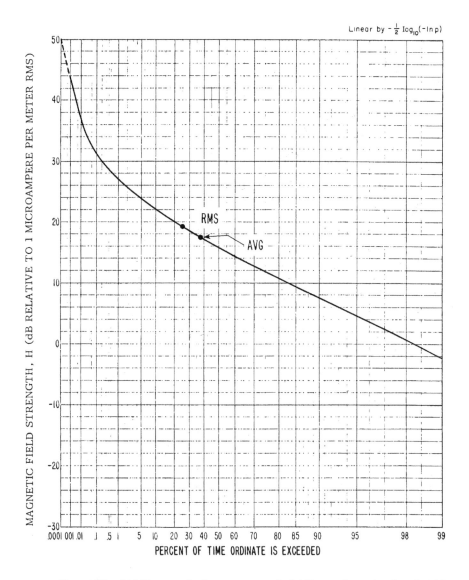

Figure 73   APD, 10 kHz, vertical component, 1.0 kHz predetection bandwidth,
December 7, 1972, 10: 45 a.m., Robena No. 4.

121

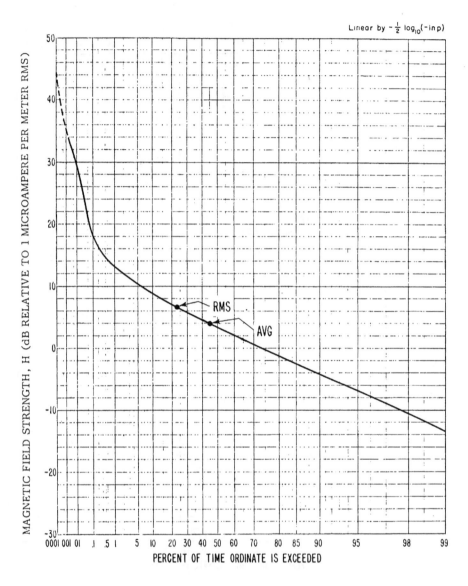

Figure 74   APD, 30 kHz, vertical component, 1.0 kHz predetection bandwidth, December 7, 1972, 11:20 a.m., Robena No. 4.

122

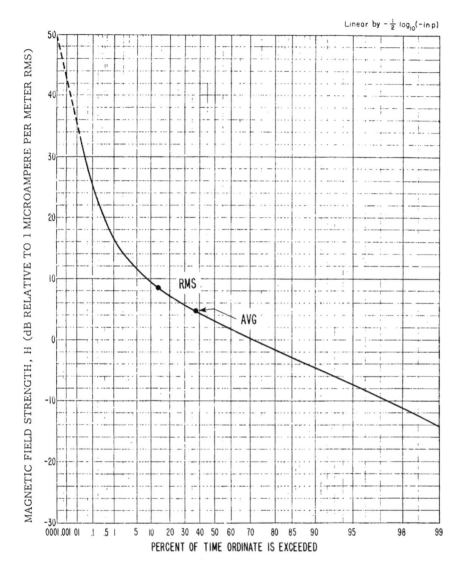

Figure 75   APD, 70 kHz, vertical component, 1.0 kHz predetection bandwidth, December 7, 1972, 11: 55 a.m., Robena No. 4.

123

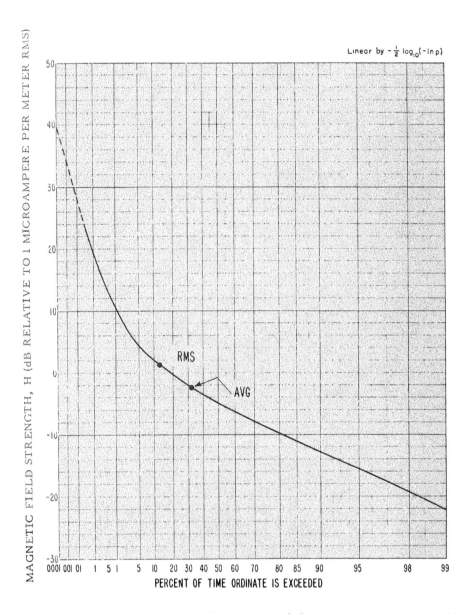

Figure 76   APD, 130 kHz, vertical component, 1.0 kHz predetection bandwidth,
December 7, 1972, 1:25 p.m., Robena No. 4.

124

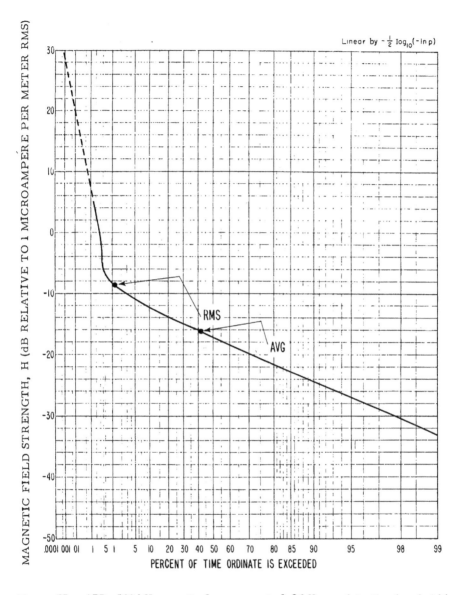

Figure 77   APD, 500 kHz, vertical component, 1.2 kHz predetection bandwidth,
December 7, 1972, 10:45 a.m., Robena No. 4.

125

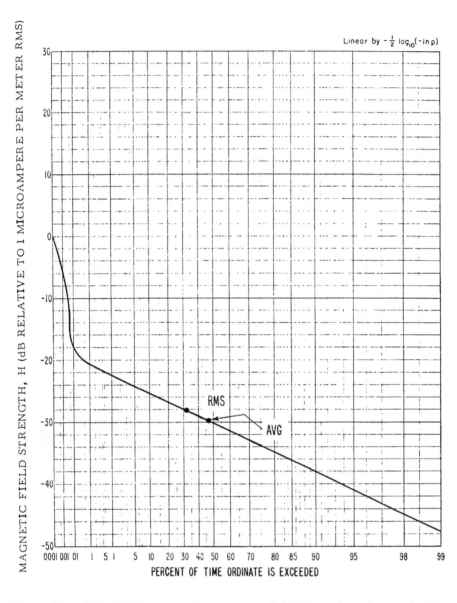

Figure 78   APD, 2 MHz, vertical component, 1.2 kHz predetection bandwidth,
December 7, 1972, 11:20 a.m., Robena No. 4.

126

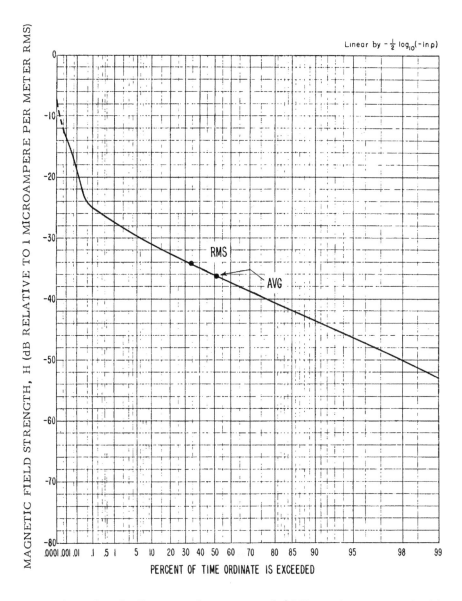

Figure 79   APD, 8 MHz, vertical component, 1.2 kHz predetection bandwidth, December 7, 1972, 11:55 a.m., Robena No. 4.

127

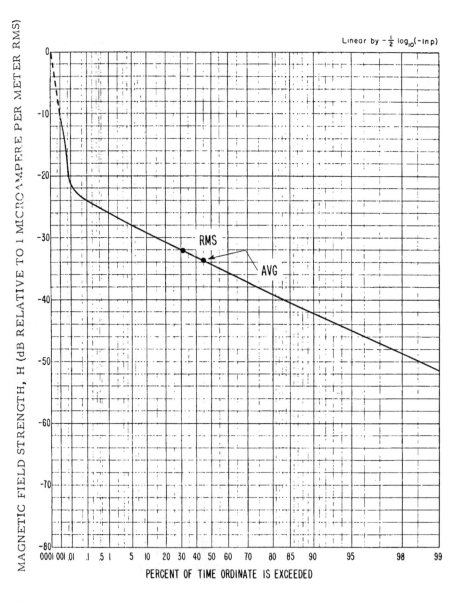

Figure 80   APD, 32 MHz, vertical component, 1.2 kHz predetection bandwidth, December 7, 1972, 1:25 p.m., Robena No. 4.

128

percentage of time the indicated level is exceeded. For the recording time interval, one can see what percentage of time a particular magnetic field strength was exceeded. In addition, one can easily infer the relative composition of the noise, i.e., whether it is Gaussian, impulsive, or CW. The Rayleigh distributed envelope of Gaussian noise has a slope of -1/2. CW noise gives slopes greater (more positive) than -1/2. Impulsive noise shows up as slopes of -4 or -5 or even more negative [9]. The noise sources and transmission effects should be considered in this respect. The impulsive noise sources are typically trolley arcs and brush arcs. At the lower frequencies, 10 kHz to 130 kHz, all wires, cables, and rails serve as relatively low-loss transmission lines, and noise generated in this part of the spectrum anywhere in the mine is transmitted throughout the mine. At the higher frequencies, the transmission loss is higher, and unless there is a local source of impulsive noise, the measured noise is Gaussian except for a relatively small percentage of the time. Above 2 MHz, for the location of these recordings (beside a trolley line but away from equipment except for occasional passing trolleys), impulsive noise is present less than one percent of the time. Harmonic power of the periodic power-line signal falls off above 10 kHz; this can be seen in the spectral plots.

129

The APD's are integrated to give rms and average values
of the field strength, according to the equations

$$H_{avg} = - \int_{0}^{\infty} H \, dp(H) \tag{2}$$

and

$$H_{rms} = \left[-\int_{0}^{\infty} H^2 \, dp(H) \right]^{\frac{1}{2}}, \tag{3}$$

where H represents the magnetic field strength of the noise,
and p is the probability that the measured field strength
exceeds the value H. These quantities (i.e., the APD quantities) are also dependent upon the measurement bandwidth, the
length of the data run, and possibly other parameters. Finite
series are actually used for the numerical integration. The
rms and average values so arrived at are identified on each
graph and are time averages (23 minutes) of these time-dependent
parameters. If the tapes are played into an rms detector, the
readings will vary 10 to 20 dB over fractions of a second.
The rms value is directly relatable to noise power. With
these wide variations of field strength with time, the most
suitable presentations are statistical ones.

Excursions of field strength between 0.1 and 99 percent,
as well as rms and average values, are shown in figures 81
through 84 for three orthogonal field components measured on
December 5, and for the vertical component measured on
December 7, 1972. The predetection bandwidth for these APD

130

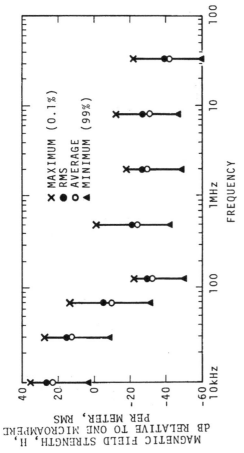

Figure 81   Field strength excursions between 0.1% and 99%
             of the time as a function of frequency, vertical
             component, December 5, 1973.

131

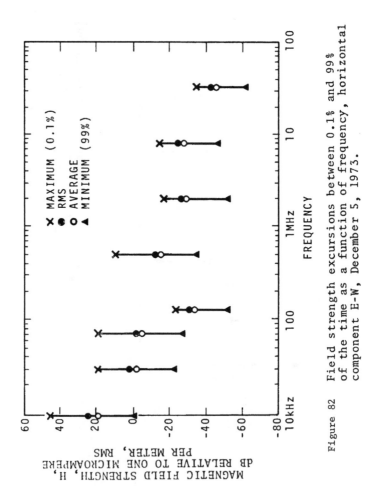

Figure 82   Field strength excursions between 0.1% and 99% of the time as a function of frequency, horizontal component E-W, December 5, 1973.

132

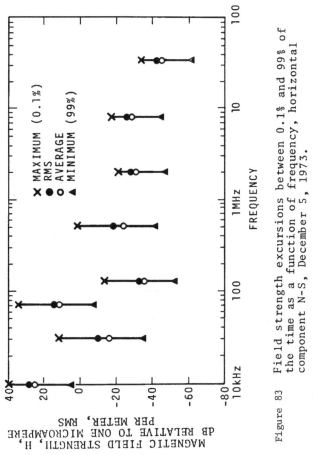

Figure 83   Field strength excursions between 0.1% and 99% of the time as a function of frequency, horizontal component N-S, December 5, 1973.

133

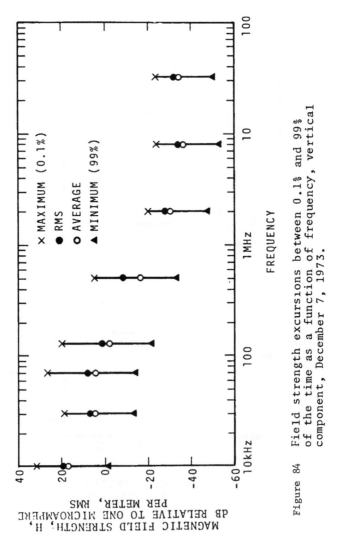

Figure 84 Field strength excursions between 0.1% and 99% of the time as a function of frequency, vertical component, December 7, 1973.

134

measurements is normalized to be 1 kHz. Some fluctuations in
values occur because of different operating conditions during
different times of the day. Between 2:00 p.m. and 4:00 p.m.
the mining equipment operated less often, but trolleys
operated more often. Considerably lower level of EM noise
around 130 kHz was observed for all three orthogonal field
components on December 5, 1972. This peculiarity was not observed on December 7, 1972, nor was it observed on several
broad-band spectral plots taken in some other mines at other
times. Therefore we conclude that the considerably lower
level of EM noise around 130 kHz on December 5 was due to
some specific operation of this coal mine on this particular
day and is not considered to be a general nature of EM noise
in a coal mine. The 23 minute time period for each measurement is adequate for covering the variations due to the local
work cycle. Shorter periods display shifts of several
decibels in APD's; longer periods (46 minutes) do not. The
emphasis in this program was to obtain noise measurements
during normal operation. We did not make measurements when
the mine was shut down. When shut down, the mine is quieter
by many tens of decibels.

IV. SPECIAL MEASUREMENTS

A. Surface Noise Measurements

1. Introduction

Surface noise measurements were made on December 5 and 7, 1972, near Blaker and Bailey Shafts of Robena No. 4 Coal Mine. The Blaker Shaft data are probably typical of farming and pasture countryside. The Bailey Shaft data are typical of mine entrances near power substations and heavy electrical machinery.

These data are relevant to electromagnetic techniques for locating entrapped miners and for special, through-the-earth, communication systems. The upper frequency limit of interest is probably no higher than 10 kHz for most practical uses. This is because attenuation through the earth increases rapidly above some "corner" frequency, often much lower than 10 kHz.

There are seven sections to part IV of this report. They are brief because the measurement techniques, instrumentation, and calibration are similar to those discussed in much more detail in other sections of the report.

2. Measurement Techniques

a. Spectra

The surface spectral measurements are made the same way as the underground spectral measurements are made, but with four differences: first, different equipment is used; second, the frequency range is less, 100 Hz to 10 kHz; third, the spectral resolution is correspondingly smaller; and fourth, a laboratory tape recorder is used, thus eliminating the need for transcribing data.

b. Amplitude Probability Distributions

Again, the measurement technique is similar to that used in making underground measurements. The differences are the same as the first and fourth listed above. The spot frequencies covered are 10 kHz, 30 kHz, ~70 kHz, and 130 kHz. These frequencies are selected for specific underground-surface measurement comparisons; the general information of this type is available elsewhere [14].

3. Measurement Instrumentation

The instrumentation used is shown in figure 85. A single laboratory tape recorder is used for both spectral and amplitude probability distribution (APD) techniques. The peripheral equipment is identical to that used in the underground system except active filters are used for the spectral

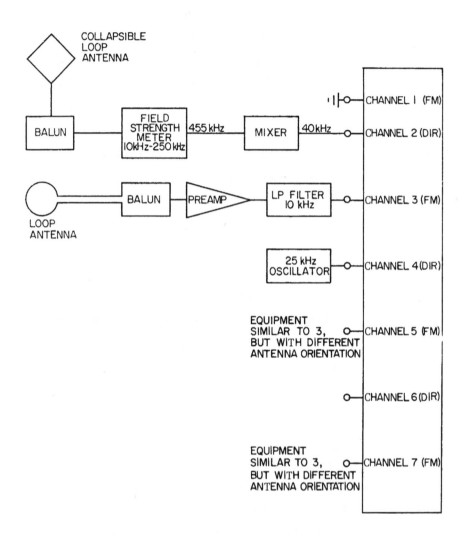

Figure 85   Surface recording system.

138

systems rather than passive filters. The size, weight, and power limitations to be considered are only moderate, and mine permissibility is not a requirement. The use of the laboratory recorder with its excellent speed-control servo system allows omission of one step, the data transcription. Also, the small frequency range of the spectral measurements requires only a four-to-one tape speed reduction for digitizing. The record speed is 15 ips, with a 10-kHz bandwidth using FM mode; reproduce speed is 3-3/4 ips, with a corresponding 2.5 kHz bandwidth.

The APD data must be recorded at the same speed, 15 ips; direct record mode is used, and, for this recorder, the 75-kHz bandwidth is quite adequate to record the 40-kHz output of the mixers. Reproduce speed is 15 ips into the same data processing system used for the underground data processing.

4. Location of Measurements

The location of measurements near Blaker Shaft is shown in figure 86. Although a power line (not shown) to a local farm house passes within 20 meters of the measurement location, and although a high tension line is about 88 meters away, there is no heavy electrical machinery and no power substation within about 229 meters. Thus the noise levels may be considered typical for dairy farming countryside.

139

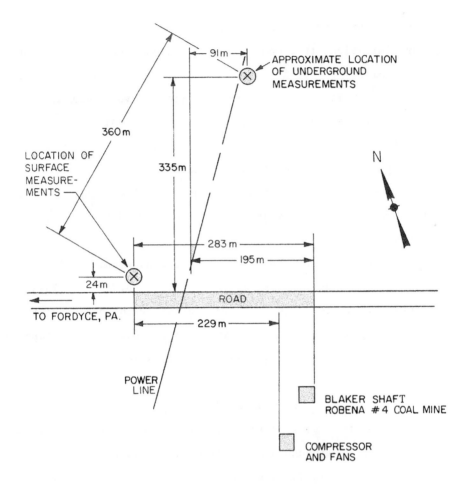

Figure 86   Location of surface measurements.

140

The original plan was to record surface data at a position directly above the position of the underground recordings, but due to sustained heavy rains before and during the date of these measurements, the overburden surface directly overhead was inaccessible. Therefore, the measurements were made with a horizontal offset of approximately 360 meters.

The measurements made near Bailey Shaft are about 30 meters from a power substation (with rectifiers for supplying underground dc power), and about 50 meters from compressors and other heavy electrical machinery. Thus, these data are indicative of a "worst case" environment, or at least what normally must be expected within 50 meters of mine entrances or other areas of heavy electrical power usage.

5. Calibration

Calibration of each spectral system and each APD system was performed as described in a previous section by immersing each antenna in a standard magnetic field and following through the entire procedure of record, reproduce, process, and display, thus calibrating each complete system.

The uncertainties in the spectral plots and APD plots are the same as stated in appropriate previous sections.

6. Results of Spectral Noise Measurements

Three orthogonal components of surface H-field noise spectral plots for Blaker Shaft are shown in figures 87

141

through 89. Similarly, three orthogonal components for Bailey Shaft are shown in figures 90 through 92. The lower curve in each figure is the system noise curve. The curves have been corrected to display absolute values between 750 Hz and 10 kHz; values below this range have increased uncertainty; values above this range are not usable. The spectral resolution is 7.81 Hz for figures 87 through 92.

With the exception of some 60-Hz harmonics the Blaker noise varies between 0 dB μA/m and -20 dB μA/m. Variations are gradual, and not consistent, but sometimes there is a broad minimum between 1 kHz and 3 kHz and a slight peak near 4 kHz. The Bailey Shaft noise is much stronger; from 1 to 2 kHz it is about 30 dB μA/m; after that it falls off at approximately 50 dB per decade.

7. Amplitude Probability Distribution Results

Only four APD measurements were made near Blaker Shaft and none at Bailey Shaft. These were made at approximately the same time, at the same frequencies, and with the same antenna sensitive axes (vertical) as for measurements underground. These APD's are shown in figures 93 through 96. The comparative results show surface noise above a working mine to be about 25 dB less than underground noise within the mine, although this generalization may be grossly in error near sources of mine noise.

142

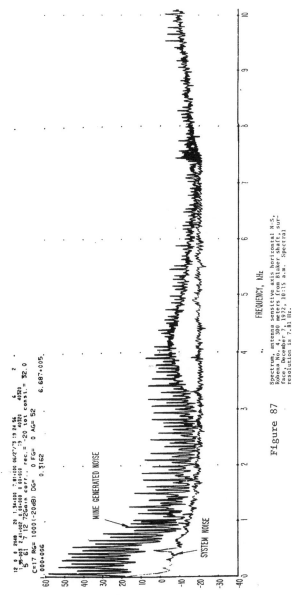

Figure 87

Spectrum, antenna sensitive axis horizontal N-S,
Robena No. 4, 300 meters from Blaker shaft, surface, December 7, 1972, 10:15 a.m. Spectral
resolution is 7.81 Hz.

143

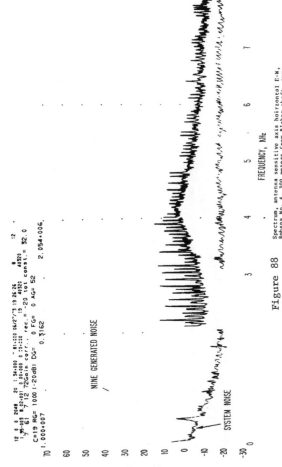

MINE GENERATED NOISE

SYSTEM NOISE

12 6 6 2048 20 1.54+000 81:z00 06/27/73 19 26 26 8 :2
1 95-003 8 02+001 0.00+000 0.z5+z00 :9 4032C 4032C
7 61 7 12 72Gain corr. rec.= -20 tot const.= 32.0

C=19 RG= 1000 (-20dB) DG= 0 FG= 0 AG= 52
1.000+007 0.3162 2.054+006.

RMS MAGNETIC FIELD STRENGTH, H, dB RELATIVE TO ONE MICROAMPERE
PER METER, FOR DISCRETE FREQUENCIES, OR
RMS MAGNETIC-FIELD-STRENGTH SPECTRUM DENSITY LEVEL, H, dB RELATIVE TO
ONE MICROAMPERE-PER-METER PER $\sqrt{T81\,Hz}$, FOR BROAD BAND NOISE

FREQUENCY, kHz

Figure 88

Spectrum, antenna sensitive axis hoirzontal E-W,
Robena No. 4, 300 meters from Blake shaft, surface, December 7, 1972, 10:15 a.m. Spectral
resolution is 7.81 Hz.

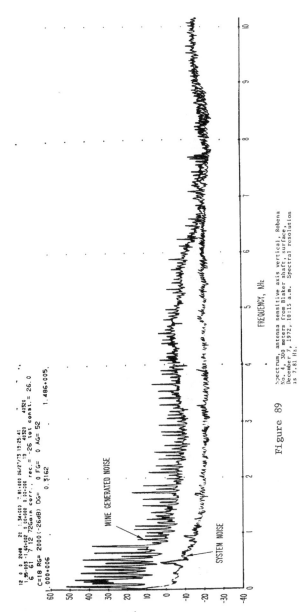

Figure 89

Spectrum, antenna sensitive axis vertical, Robena
No. 4, 300 meters from Blaker shaft, surface,
December 7, 1972, 10:15 a.m. Spectral resolution
is 7.81 Hz.

145

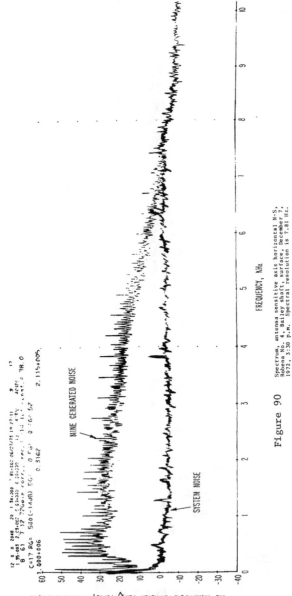

Figure 90   Spectrum, antenna sensitive axis horizontal N-S,
            Robena No. 4, Bailey shaft, surface, December 7
            1972, 3:30 p.m. Spectral resolution is 7.81 Hz.

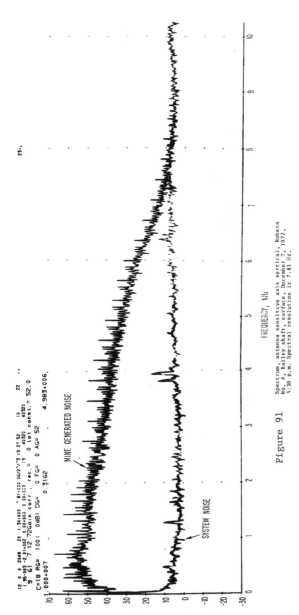

Figure 91

Spectrum, antenna sensitive axis vertical, Robena
No. 4, Bailey shaft, surface, December 7, 1972,
4:30 p.m. Spectral resolution is 7.81 Hz.

147

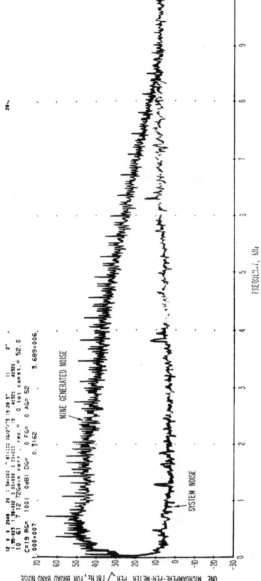

Figure 92

Spectrum, antenna sensitive axis horizontal E-W.
Robena No. 4, Bailey shaft, surface, December 7,
1972, 3 30 p.m. Spectral resolution is 7.81 Hz.

148

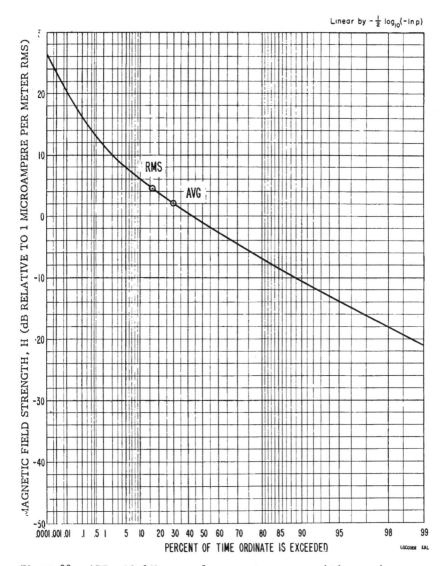

Figure 93   APD, 10 kHz, surface, antenna sensitive axis
vertical, Robena No. 4, December 5, 1972,
11:45 a.m.

149

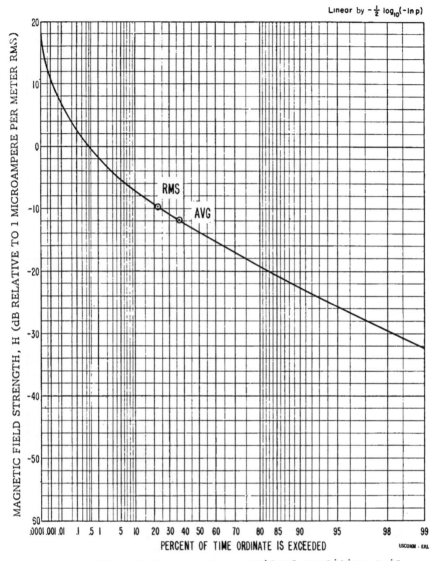

Figure 94   APD, 30 kHz, surface, antenna sensitive axis
vertical, Robena No. 4, December 5, 1972,
2:00 p.m.

150

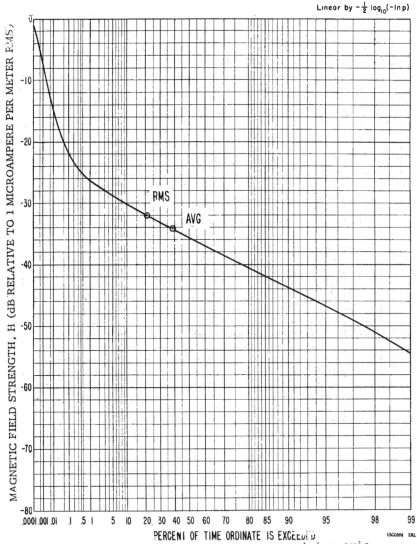

Figure 95   APD, 70 kHz, surface, antenna sensitive axis
vertical, Robena No. 4, December 5, 1972,
3:45 p.m.

151

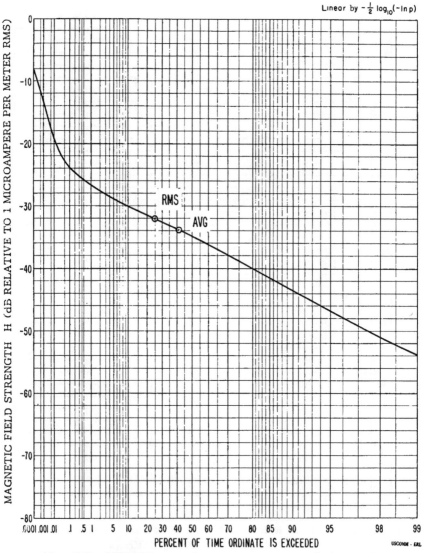

Figure 96   APD, 130 kHz, surface, antenna sensitive axis
vertical, Robena No. 4, December 5, 1972,
5:15 p.m.

152

At 10 kHz the surface noise varies between 26 dB μA/m
(0.0001% of time) and -20 dB μA/m (99% of time).

A summary of the field strength excursions between 0.1%
and 99% of the time as a function of frequency is shown in
figure 97 for the surface data.

B. Measurements of Voltage Between "Roof Bolts"

Measurements of RF voltage were made between three pairs
of roof bolts with intrapair spacings of 3.6 m, 8.5 m, and
13 m. Roof-bolt locations are shown in figure 5. Unshielded,
insulated copper wire is attached to the bolts using large
clips. The other end of the wire pair is connected directly
to the input of an amplifier with 10^8 ohms input impedance.
The resulting voltage spectra are shown in figures 98, 99,
and 100, for spacing of 3.6, 8.5, and 13 meters, respectively.
The logarithmic average of the 360 Hz harmonics 3 through 8
for the 13-meter spacing is about -84.2 dB with respect to
1 V rms, i.e., 62 μV rms. The amplitudes in the spectra for
3.6- and 8.5-meter spacings are too low to allow reliable
scaling of voltages.

Figure 101 shows the voltage spectrum obtained from an
8.5-meter "dipole". Instead of being clipped onto the roof
bolts, the wires are merely left unconnected. The resulting
voltage is much higher. The logarithmic average of harmonics

153

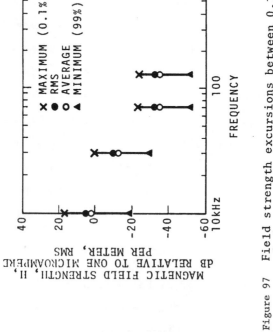

Figure 97   Field strength excursions between 0.1% and 99% of the time as a function of frequency, surface vertical component, December 5, 1973.

154

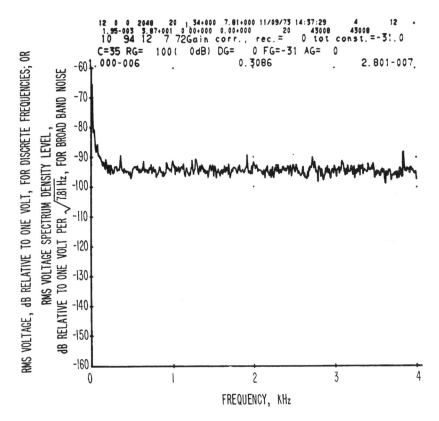

Figure 98   Voltage spectrum, 100 Hz to 3 kHz, Robena No. 4
mine, underground, roof bolts-3.6-meter separation, 12:25 p.m., Dec. 7, 1972. Spectral resolution is 7.81 Hz.

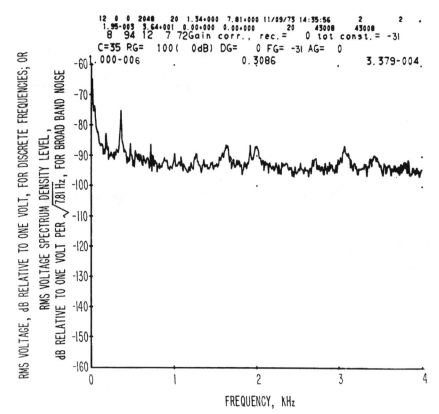

RMS VOLTAGE, dB RELATIVE TO ONE VOLT, FOR DISCRETE FREQUENCIES; OR

RMS VOLTAGE SPECTRUM DENSITY LEVEL, dB RELATIVE TO ONE VOLT PER $\sqrt{7.81\,Hz}$, FOR BROAD BAND NOISE

FREQUENCY, KHz

Figure 99   Voltage spectrum, 1 3 Hz to 5 kHz, Robena No. 4 mine, underground, roof bolts-8.5-meter separation, 12:20 p.m., Dec. 7, 1972, Spectral resolution is 7.81 Hz.

156

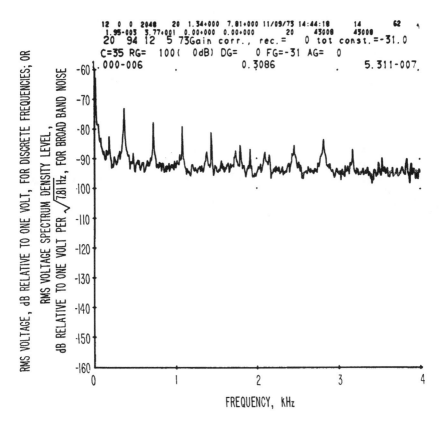

RMS VOLTAGE, dB RELATIVE TO ONE VOLT, FOR DISCRETE FREQUENCIES; OR

RMS VOLTAGE SPECTRUM DENSITY LEVEL, dB RELATIVE TO ONE VOLT PER $\sqrt{7.81\text{Hz}}$, FOR BROAD BAND NOISE

FREQUENCY, kHz

Figure 100   Voltage spectrum, 100 Hz to 3 kHz, Robena No. 4
mine, underground, roof bolts -- 13-meter separation, 12:14 p.m., Dec. 7, 1972.   Spectral resolution is 7.81 Hz.

157

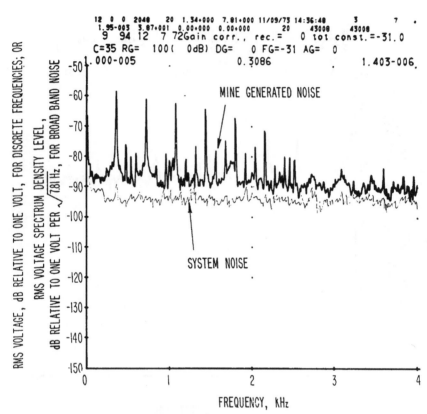

Figure 101 Voltage spectrum, 100 Hz to 3 kHz, Robena No. 4
mine, underground, dipole antenna, 8.5-meters
long, 12:24 p.m., Dec. 7, 1972. Spectral resolution is 7.81 Hz.

3 through 8 is 70.0 dB below 1 volt, i.e., 316 μV rms (14.2 dB higher than the 13-meter roof bolt voltage).

Data taken simultaneously on a pair of roof bolts and on a loop antenna show unlike responses. At one time, a particular 88-kHz signal would be picked up more strongly by the roof bolts than by the loop. The same is true for impulses. To illustrate this, figure 102 shows the signal received on a loop antenna at position 2 of figure 5. Figure 103 shows the signal taken simultaneously on a pair of roof bolts separated by 13 meters. A fairly strong impulse is shown on the bolts, while this impulse does not appear above the loop system noise. These differences in antenna pick-up characteristics should be studied in more detail.

C. Surface-Underground Noise Coherence Tests

A test was made of the coherence between noise at the surface and noise underground at Bailey shaft. Direct current power lines are associated with the shaft. Very large mine supply currents, both underground and on the surface, are present near the receiving antennas. As a result, the measurement is probably more one of noise coherence along a 600 foot run of dc supply cable rather than of the coherence of atmospheric or other noise directly between surface and underground.

A system test was performed by injecting a single broadband noise source into the two channels used for the coherence

159

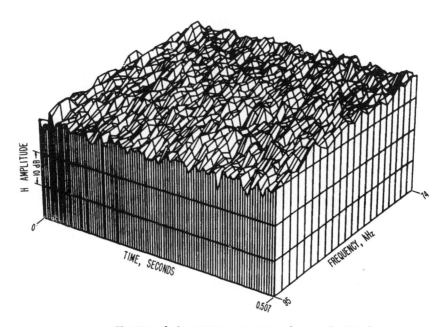

Figure 102   3D plot of the voltage spectrum of magnetic field
strength obtained on a loop antenna, 74 kHz to
95 kHz, Robena No. 4 mine, underground,
12:14 p.m., Dec. 5, 1972, antenna sensitive axis
vertical, no impulse is detectable. Spectral
resolution is 1000 Hz.

160

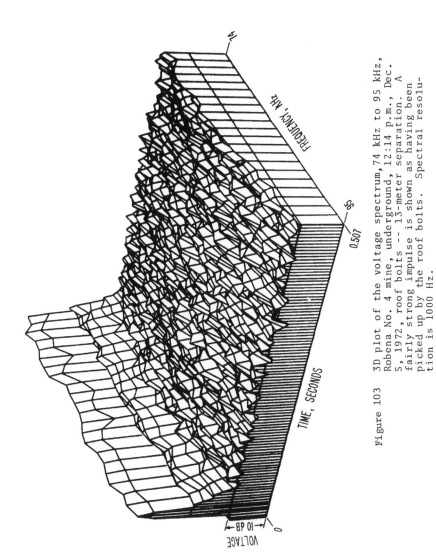

Figure 103   3D plot of the voltage spectrum, 74 kHz to 95 kHz, Robena No. 4 mine, underground, 12:14 p.m., Dec. 5, 1972, roof bolts -- 13-meter separation. A fairly strong impulse is shown as having been picked up by the roof bolts. Spectral resolution is 1000 Hz.

161

test. One channel required the use of 800 ft. (245 m) of RG-58 cable. Test results show that the cables and tape-recorder flutter introduced negligible coherence degradation over the frequency band of interest.

Figure 104 shows the results of the coherence measurements at Bailey Shaft. The coherence is above 0.85 in the frequency range 270 Hz to 6.6 kHz. For this frequency range and for these conditions, variations in noise measured on the surface would be very similar to noise variations made underground. The coherence as defined by Benignus [15] is:

$$\hat{\gamma}^2(F) = \frac{\hat{G}_{xy}{}^2(F)}{\hat{G}_{xx}(F) \cdot \hat{G}_{yy}(F)}$$

where $\hat{G}_{xy}(F)$ is the cross power spectrum estimate, $\hat{G}_{xx}(F)$ is the power spectrum estimate of time series x, $\hat{G}_{yy}(F)$ is the power spectrum estimate of time series y, and F is the frequency index.

D. Mine Phone Wire Measurements

1. Voltage

Measurements on mine phone-wire voltage relative to the rail conductor were made in the working section at location numbered 2 in figure 5. Figures 105 and 106 show the spectrum obtained December 5, 1972, during second-shift operation. The

162

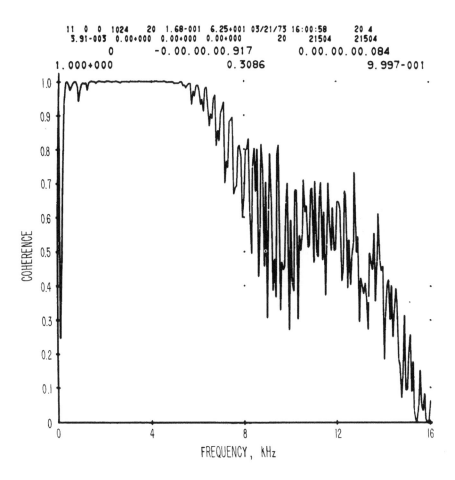

11 0 0 1024   20 1.68-001 6.25+001 03/21/73 16:00:58   20 4
3.91-003 0.00+000 0.00+000 0.00+000   20   21504   21504
0   -0.00.00.00.917   0.00.00.00.084
1.000+000   0.3086   9.997-001

Figure 104   Spectrum of the coherence of noise received on the
surface and underground, Robena No. 4 mine,
4:00 p.m., Dec. 7, 1972. Data spectra used to
generate this spectrum were 300 Hz to 40 kHz with
a spectral resolution of 62.5 Hz.

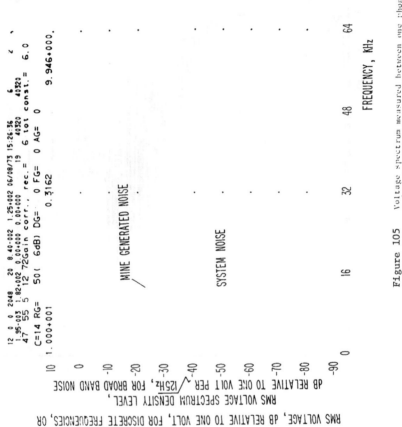

Figure 105   Voltage spectrum measured between one phone wire
(of two) relative to a rail, 1 kHz to 100 kHz,
Rohena No. 4 mine, underground, 4:30 p.m., Dec. 5
1972. Spectral resolution is 125 Hz.

164

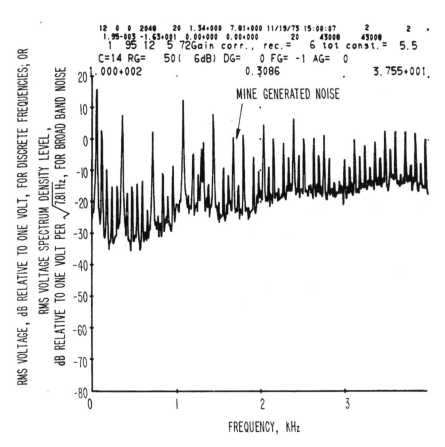

RMS VOLTAGE, dB RELATIVE TO ONE VOLT, FOR DISCRETE FREQUENCIES; OR

RMS VOLTAGE SPECTRUM DENSITY LEVEL, dB RELATIVE TO ONE VOLT PER $\sqrt{7.81 Hz}$, FOR BROAD BAND NOISE

12 0 0 2048   20 1.34+000 7.81+000 11/19/73 15:00:07   2   2 .
1.95-003 -1.63+001 0.00+000 0.00+000   20   43008   43008
1   95 12   5 72Gain corr., rec. =   6 tot const. = 5.5
C=14 RG=   50(6dB) DG=   0 FG= -1 AG=   0
1.000+002   0.3086   3.755+001.

MINE GENERATED NOISE

FREQUENCY, kHz

Figure 106   Spectrum of the voltage measured on the mine phone
wire relative to the rail, Robena No. 4 mine,
underground, 100 Hz to 3 kHz, 4:30 p.m., Dec. 5,
1972. Spectral resolution is 7.81 Hz.

165

logarithmic average of harmonics 3 through 8 in figure 106 is 5 dB relative to 1 V rms, i.e., 1.8 V rms. An unusual feature of the phone line voltage spectrum is the very high amplitude at 60 Hz as shown on figure 106.

2. Measured Noise Current

Figures 107 and 108 show the noise current in one mine phone wire, measured at location 2 (figure 5) on December 7, 1972. The signals in the audio region do not appear to be 360 Hz harmonics. The current appears to be very low, and the logarithmic average in the region of 360 Hz harmonics 3 through 8 is -92.8 dB relative to one ampere rms, i.e. 23 µA rms. The current at 88 kHz peaks at about -80 dB, i.e., 100 µA rms.

E. Trolley Wire Voltage Measurements

Figures 109 and 110 show the spectrum of the voltage measured on the 600 volt dc trolley wire relative to the rail in the working section at 4:59 p.m. on December 5, 1972. The logarithmic average of harmonics 3 through 8 of 360 Hz is +10.1 dB relative to 1 volt rms, i.e., 3.2 volts rms. The amplitude of the 88-kHz mine phone FM signal is 29 dB above a volt, i.e., 28.2 volts rms. Variations in harmonic voltages are seen by comparing figures 109 and 111, taken two days apart. In figure 111, logarithmic average of harmonics 3 through 8 is -10.2 dB, i.e. 0.31 V rms. The 88-kHz mine phone carrier was measured to be 100 volts rms.

166

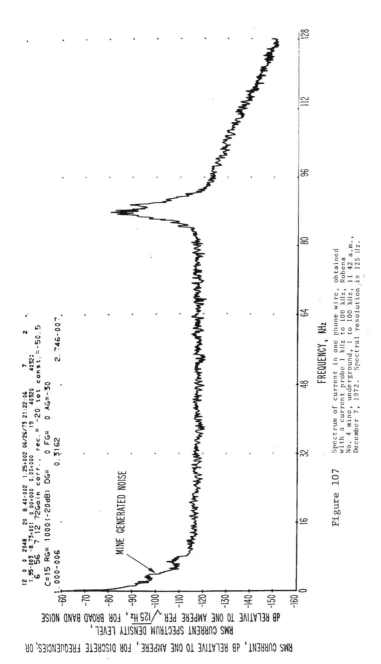

Figure 107   Spectrum of current in one phone wire, obtained
with a current probe 1 kHz to 100 kHz, Robena
No. 4 mine, underground, 1 to 100 kHz, 11 42 a.m.,
December 7, 1972. Spectral resolution is 125 Hz.

167

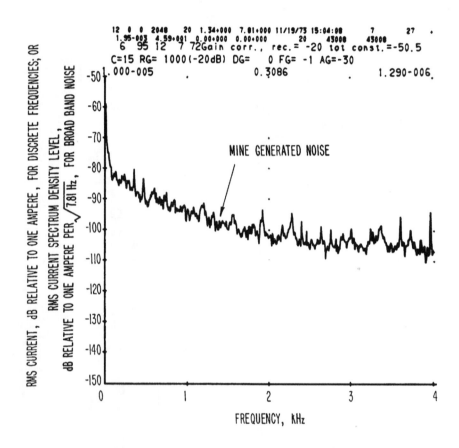

Figure 108   Spectrum of the current on one phone wire obtained
with a current probe, Robena No. 4 mine, underground, 100 Hz to 3 kHz, 11:42 a.m., Dec. 7, 1972.
Spectral resolution is 7.81 Hz.

168

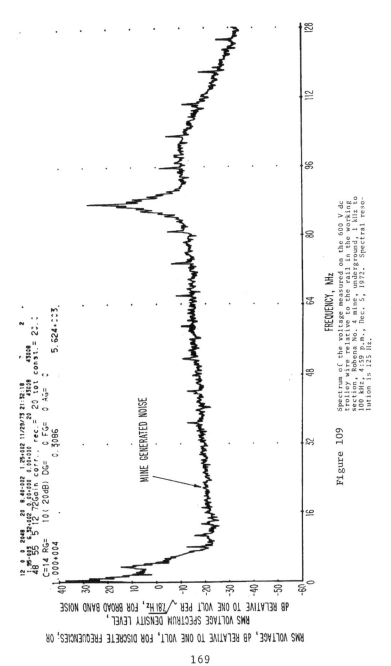

Figure 109   Spectrum of the voltage measured on the 600 V dc trolley wire relative to the rail in the working section, Robena No. 4 mine, underground, 1 kHz to 100 kHz, 4:59 p.m., Dec. 5, 1972. Spectral resolution is 125 Hz.

169

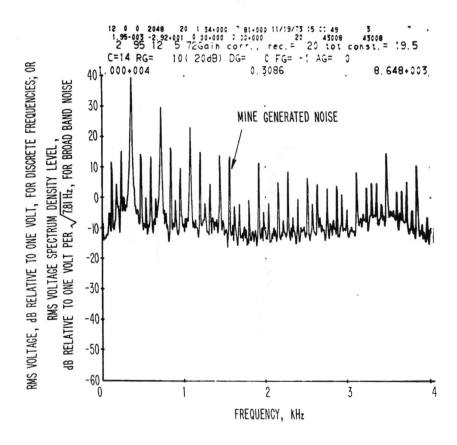

RMS VOLTAGE, dB RELATIVE TO ONE VOLT, FOR DISCRETE FREQUENCIES; OR
RMS VOLTAGE SPECTRUM DENSITY LEVEL,
dB RELATIVE TO ONE VOLT PER $\sqrt{7.81\,Hz}$, FOR BROAD BAND NOISE

FREQUENCY, kHz

Figure 110   Spectrum of the voltage measured on the 600 V dc
trolley wire relative to the rail in the working
section, 100 Hz to 3 kHz, Robena No. 4 mine,
underground, 100 Hz to 3 kHz, 4:59 p.m., Dec. 5,
1972. Spectral resolution is 7.81 Hz.

170

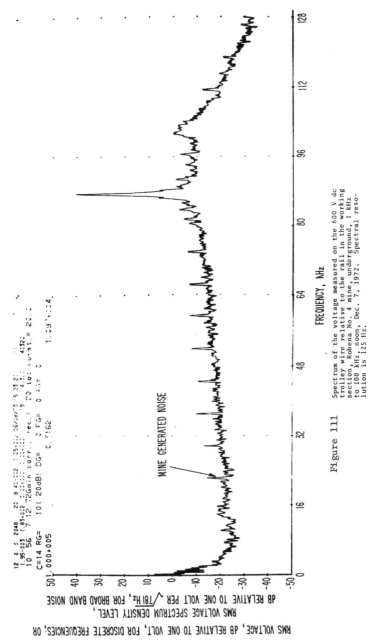

Figure 111   Spectrum of the voltage measured on the 600 V dc
trolley wire relative to the rail in the working
section, Robena No. 4 mine, underground, 1 kHz
to 100 kHz, noon, Dec. 7, 1972. Spectral resolution is 125 Hz.

171

V. CONCLUSIONS

1.   The amplitude of magnetic field noise as a function
of frequency is shown in the spectral plots of section II, B.
In most cases, the noise energy in this mine occurs in the
spectrum below 10 kHz and commonly is 50 to 100 dB above 1
microampere per meter in this noisy portion of the spectrum
when measured near (within about 3 meters) equipment or power
cables.

2.   Above 20 kHz, the magnetic noise is generally within
-20 to +30 dB with respect to one microampere per meter,
except at trolley-phone carrier frequencies. It may be considerably lower in some locations.

3.   The 88 or 100 kHz trolley phones, when voice modu-
lated, cover ± 10 kHz. In some cases, they create harmonics
up to 8 MHz.

4.   Moving 30 meters away from machinery and power lines
causes a significant drop in noise, up to 40 dB.

5.   The APD's show a 40 to 50 dB variation of noise with
time, with impulsive noise present up to 30 percent of the time
at frequencies below 1 MHz, while the impulsive noise is
usually present less than one percent of the time at frequencies
above 1 MHz.

6.   The APD's also show that there is a general decrease
in noise as frequencies increase, ranging in this mine from
+20 dB µA/m at 10 kHz to -40 dB µA/m at 30 MHz. These are
time-averaged rms values.

172

7. Magnetic field noise measured on the surface above a mine depends on proximity to surface noise sources, but is, in general, less than underground noise, at least in the winter. It varied from +10 dB μA/m at 10 kHz to -30 dB μA/m at 130 kHz at one location about 300 meters from Blaker Shaft.

8. Surface noise may be correlated to underground noise over a portion of the spectrum dominated by power line harmonics; at Robena Mine this is below 10 kHz. At higher frequencies, there is no correlation.

9. Trolley wire-to-rail noise voltages have spectral contents similar to H-field results; peak amplitudes are at 360 Hz (approximately 100 volts rms) and other power line harmonics. Transients sometimes exceed these peak values.

10. Roof-bolt voltages also have spectral contents similar to H-field spectra; they may be more sensitive to impulsive noise in some cases than loop antennas are, but the reverse was also observed. Further investigation is needed.

VI. RECOMMENDATIONS

Electromagnetic noise is a dominant factor in range-limited communication systems. In mine communication systems, range-limiting factors are high attenuation, low size-weight considerations, cost considerations, power-permissibility limitations, and others. In addition, the in-mine environment is much noisier than most other environments, and has more different noise sources of different types than most other environments. Therefore, the character and magnitude of this electromagnetic noise must be known in order to be overcome.

Our principal recommendation is to use the data and techniques discussed in this report to study the character and magnitude of electromagnetic noise in mine environments. It should be recognized that there is no one interpretation or solution.

In the distant future, improved equipment design may reduce the noise.

VII. ACKNOWLEDGMENTS

Those making significant contributions to this program are as follows: laboratory development and field use of measurement equipment, Ed Neisen, Doug Schulze, and Tom Bremer; data processing, Ann Rumfelt, Nancy Tomoeda, and Frank Cowley. Those making valuable but less time-consuming contributions are Gerry Reeve, Bob Matheson, Don Spaulding, John Chukoski, Lorne Matheson, and Dave Lewis.

Winston Scott and Don Halford provided much assistance in proofreading, while Sharon Foote and Janet Becker typed tirelessly through many versions. Jocelyn Spencer provided drafting assistance.

Finally, none of this would have been possible without the excellent cooperation of Bob Goddard, Bill Zeller, and others at Robena No. 4 Mine of United States Steel Corporation.

VIII. REFERENCES

[1] The Institute of Electrical and Electronic Engineers, Inc., IEEE Dictionary of Electrical and Electronic Terms, Std. 100, 1972.

[2] Electric Motor-Driven Mine Equipment and Accessories, Schedule 2G, Federal Register, Vol. 23, No. 54, March 19, 1968.

[3] Corbin, C.S., Servo from Tape, Honeywell Technical Talk, Denver (MS-211), July 23, 1971.

[4] Burnett, E.D., Corliss, E.L.R., and Berendt, R.D., Magnetic recording of acoustic data on audiofrequency tape, NBS Technical Note 718, April 1972.

[5] Welch, P.D., The use of fast Fourier transform for the estimation of power spectra. A method based on time averages over short, modified, periodograms, IEEE Trans. on Audio and Electroacoustics, Vol. AU-15, No. 2, June 1967, pp. 70-73.

[6] Greene, F.M., NBS field-strength standards and measurements (30 Hz to 1000 MHz), Proc. IEEE, Vol. 55, No. 6, June 1967, pp. 970-981.

[7] Crichlow, W.Q., et al., Amplitude-probability distributions for atmospheric radio noise, NBS Monograph 23 (1960b).

[8] Thompson, W.I., III, Bibliography of ground vehicle communications and control. AKWIC index, Report No. DOT-TSC-UMTA-71-3, July 1971.

9] Shaver, Harry V.; V. Elaine Hatfield, George H. Hugh,
Man-made radio noise parameter identification task,
Final Report--Naval Electronics Systems Command, Stand-
ard Research Report, SRI Project 1022-2, May 1972.

10] Matheson, R.J., Instrumentation problems encountered
making man-made electromagnetic noise measurements for
predicting communication system performance, IEEE Trans.
on EM compatibility, Vol. EMC-12, No. 4, November 1970,
pp. 151-158.

11] Greene, F.M., Calibration of commercial radio field strength
meters at the National Bureau of Standards, NBS Circular
517, 1951.

12] Jean, A.G., Taggart, H.E., and Wait, J.R., Calibration
of loop antennas at VLF, NBS J. Res., 64, No. 3, 1961.

3] Greene, F.M., The near-zone magnetic field of a small
circular-loop antenna, NBS J. Res., 71C, No. 4, Oct-
Dec. 1967.

4] C.C.I.R. Report 322, World distribution and characteristics
of atmospheric radio noise, Documents of the Xth Plenary
Assembly, I.T.U., (1964).

5] Benignus, V.A., Estimation of the coherence spectrum and
its confidence interval using the Fast Fourier Transform,
IEEE Trans. on Audio and Electroacoustics, Vol. AU-17,
No. 2, June 1969, pp. 145-150.

Errata - NBS Technical Note 654

gures 87 through 92, substitute "surface noise" for "mine
nerated noise."

Some additional uncertainty beyond the stated measureme
system uncertainty is caused by the in-mine environment. Ca
was taken to provide at least one meter separation from
metallic objects wherever possible. However, coal, rock, or
earth was sometimes immediately adjacent to a loop antenna.
In all observed cases, this had no effect at frequencies up
to 1 MHz. Above 1 MHz, earth and reflections did in some
cases cause ± 1 dB variations, even with a shielded, balance
loop antenna. An estimate is that an additional ± 5 dB unce
tainty might be advisable. However, due to the complexity o
the shielded loop in the mine environment, this uncertainty
cannot be rigorously bounded without substantial additional
analysis.

Decoding of Spectrum Captions

Spectrum captions are generally organized into the following format:

First line: MP NDT NZS NDA NPO RC DF date, time, frame, serial,
where

MP   = Two's power of length of Fourier transform, example,
       2^{MP} where MP = 11

NDT = Detrending option, example, 0 (dc removed)

NZS = Restart spectral average after output, example, 0
       (restarted)

NDA = Data segment advance increment, example, 2048

NPO = Number of spectra averaged between output calls,
       example, 20

RC   = Integration time in seconds per spectra, example, 0.168

DF   = Resolution bandwidth, spectral estimate spacing in
       hertz, example, 62.5

Date = Date of computer processing, example, 03/21/73

Time = Time of computer processing, example, 15:06:34

Frame= Frame set number, example, 10

Serial = Film frame serial number, example, 42.

Second line: DTA DA(1) DA(2) DA(3) NSA NRP NPP, where

DTA   = Detrending filter parameter α, example, 0.00195

DA(1) = Detrending filter average, K=1, example, 59.4

177

DA(2) = Detrending filter average, K=2, example, 0

DA(3) = Detrending filter average, K=3, example, 0

NSA = Number of periodograms averaged, example, 20

NRP = Number of data points processed since spectrum initialization, example, 43008

NPP = Number of data points processed since data initialization, example, 43008.

Third line: RUN, SESSION, DAY, MONTH, YEAR Gain corr., rec. = tot. constr. =, where

Run and Session = the title of the portrayed frame identifying the digitizing session and run number, example, 33 55

Day, Month, Year = date data were recorded in the mine, example, 5 12 73

Gain corr. rec. = receiver gain correction, example, 0

tot. const. = constant gain correction of entire system, example, -30.5.

Fourth line: Top of Scale, Standard Error, Spectral Peak, where

Top of Scale = largest scale marking for computer drawn graph, example, 1.000-005 (1.0×10^{-5})

Standard Error = standard error of curve, example, 0.3162

Spectral Peak = largest spectral peak observed, example, 7.747-006. (7.747×10^{-6})

The preceding coding was used extensively in the Robena report, but the format has recently been changed, and a few examples of the new format will be found in this report. In the new format, the old fourth line is now the new fifth line and the new fourth line has the following format:

New Fourth line: C =, RG =, DG =, FG =, AG =, where

C = correction curve used with data, example, 34

RG = receiver gain and accompanying correction in dB added to the data, example, 10 (20 dB)

DG = digitizer gain, example, 0

FG = filter gain in dB, often rounded to nearest single digit, example, -1

AG = absolute gain correction added to data, example, 0

lectromagnetic Noise in Robena No. 4 Coal Mine

9. PERFORMING ORGANIZATION NAME AND ADDRESS

NATIONAL BUREAU OF STANDARDS, Boulder Labs
DEPARTMENT OF COMMERCE

Boulder, Colorado 80302

12. Sponsoring Organization Name and Complete Address *(Street, City, State, ZIP)*

U. S. Bureau of Mines
Pittsburgh Mining and Safety Research Center
4800 Forbes Avenue
Pittsburgh, Pennsylvania 15213

16. ABSTRACT *(A 200-word or less factual summary of most significant information. If document includes a significant bibliography or literature survey, mention it here.)*

Two different techniques were used to make measurements of the absolute value electromagnetic noise in an operating coal mine, Robena No. 4, located near Wayne Pennsylvania. One technique measures noise over the entire electromagnetic spect interest for brief time periods. With present instrumentation, the spectrum can covered from 40 Hz to 400 kHz. It is recorded using broad-band analog magnetic t and the noise data are later transformed to give spectral plots. The other techn records noise envelopes at several discrete frequencies for a sufficient amount o to provide amplitude probability distributions.

The specific measured results are given in a number of spectral plots and amp probability distribution plots. The general results are that at frequencies belo power line noise within the mine is severe. Impulsive noise is severe near arcin trolleys, and at lower frequencies near any transmission line. Carrier trolley p signals and harmonics are strong throughout the mine whenever the trolley phone i operation.

Additional information beyond that included in this report may be obtained fr authors, who are with the Electromagnetics Division of the National Bureau of Sta

17. KEY WORDS *(six to twelve entries, alphabetical order, capitalize only the firs name, separated by semicolons)* Amplitude probability distrib data; electromagnetic interference; electromagnetic no Fast Fourier Transform; Gaussian distribution; impuls measurement instrumentation; spectral density; time-de

first key word unles .1 mine noise; ;ency communic magnetic fie ectral densit

18. AVAILABILITY   [X] Unlimited

[ ] For Official Distribution. Do Not Release to NTIS

[ ] Order From Sup. of Doc., U.S. Government Printing Office
Washington, D.C. 20402, SD Cat. No. C13, 46:654

[ ] Order From National Technical Information Service (NTIS)
Springfield, Virginia 22151

NBS TECHNICAL PUBLICATIONS

PERIODICALS

JOURNAL OF RESEARCH reports National Bureau of Standards research and development in physics, mathematics, and chemistry. Comprehensive scientific papers give complete details of the work, including laboratory data, experimental procedures, and theoretical and mathematical analyses. Illustrated with photographs, drawings, and charts Includes listings of other NBS papers as issued

Published in two sections, available separately:

● Physics and Chemistry (Section A)

Papers of interest primarily to scientists working in these fields This section covers a broad range of physical and chemical research, with major emphasis on standards of physical measurement, fundamental constants, and properties of matter Issued six times a year. Annual subscription Domestic, $17 00; Foreign, $21.25.

● Mathematical Sciences (Section B)

Studies and compilations designed mainly for the mathematician and theoretical physicist. Topics in mathematical statistics, theory of experiment design, numerical analysis, theoretical physics and chemistry, logical design and programming of computers and computer systems Short numerical tables Issued quarterly. Annual subscription. Domestic, $9.00; Foreign, $11 25

DIMENSIONS/NBS (formerly Technical News Bulletin)—This monthly magazine is published to inform scientists, engineers, businessmen, industry, teachers, students, and consumers of the latest advances in science and technology, with primary emphasis on the work at NBS.

DIMENSIONS/NBS highlights and reviews such issues as energy research, fire protection, building technology, metric conversion, pollution abatement, health and safety, and consumer product performance In addition, DIMENSIONS/NBS reports the results of Bureau programs in measurement standards and techniques, properties of matter and materials, engineering standards and services, instrumentation, and automatic data processing.

Annual subscription Domestic, $6 50, Foreign, $8 25

NONPERIODICALS

Monographs—Major contributions to the technical literature on various subjects related to the Bureau's scientific and technical activities

Handbooks—Recommended codes of engineering and industrial practice (including safety codes) developed in cooperation with interested industries, professional organizations, and regulatory bodies

Special Publications—Include proceedings of high-level national and international conferences sponsored by NBS, precision measurement and calibration volumes, NBS annual reports, and other special publications appropriate to this grouping such as wall charts and bibliographies.

Applied Mathematics Series—Mathematical tables, manuals, and studies of special interest to physicists, engineers, chemists, biologists, mathematicians, computer programmers, and others engaged in scientific and technical work

National Standard Reference Data Series—Provides quantitative data on the physical and chemical properties of materials, compiled from the world's literature and critically evaluated Developed under a world-wide program coordinated by NBS. Program under authority of National Standard Data Act (Public Law 90-396). See also Section 1.2 3

Building Science Series—Disseminates technical information developed at the Bureau on building materials, components, systems, and whole structures. The series presents research results, test methods, and performance criteria related to the structural and environmental functions and the durability and safety characteristics of building elements and systems

Technical Notes—Studies or reports which are complete in themselves but restrictive in their treatment of a subject. Analogous to monographs but not so comprehensive in scope or definitive in treatment of the subject area Often serve as a vehicle for final reports of work performed at NBS under the sponsorship of other government agencies.

Voluntary Product Standards—Developed under procedures published by the Department of Commerce in Part 10, Title 15, of the Code of Federal Regulations. The purpose of the standards is to establish nationally recognized requirements for products, and to provide all concerned interests with a basis for common understanding of the characteristics of the products. The National Bureau of Standards administers the Voluntary Product Standards program as a supplement to the activities of the private sector standardizing organizations.

Federal Information Processing Standards Publications (FIPS PUBS)—Publications in this series collectively constitute the Federal Information Processing Standards Register. The purpose of the Register is to serve as the official source of information in the Federal Government regarding standards issued by NBS pursuant to the Federal Property and Administrative Services Act of 1949 as amended, Public Law 89-306 (79 Stat. 1127), and as implemented by Executive Order 11717 (38 FR 12315, dated May 11, 1973) and Part 6 of Title 15 CFR (Code of Federal Regulations). FIPS PUBS will include approved Federal information processing standards information of general interest, and a complete index of relevant standards publications.

Consumer Information Series—Practical information, based on NBS research and experience, covering areas of interest to the consumer. Easily understandable language and illustrations provide useful background knowledge for shopping in today's technological marketplace.

NBS Interagency Reports—A special series of interim or final reports on work performed by NBS for outside sponsors (both government and non-government). In general, initial distribution is handled by the sponsor; public distribution is by the National Technical Information Service (Springfield, Va 22151) in paper copy or microfiche form.

Order NBS publications (except Bibliographic Subscription Services) from: Superintendent of Documents, Government Printing Office, Washington, D C. 20402.

BIBLIOGRAPHIC SUBSCRIPTION SERVICES

The following current-awareness and literature-survey bibliographies are issued periodically by the Bureau:

Cryogenic Data Center Current Awareness Service (Publications and Reports of Interest in Cryogenics). A literature survey issued weekly. Annual subscription: Domestic, $20 00; foreign, $25 00

Liquefied Natural Gas. A literature survey issued quarterly. Annual subscription· $20.00

Superconducting Devices and Materials. A literature survey issued quarterly. Annual subscription. $20.00 Send subscription orders and remittances for the pre-

ceding bibliographic services to the U.S. Department of Commerce, National Technical Information Service, Springfield, Va. 22151

Electromagnetic Metrology Current Awareness Service (Abstracts of Selected Articles on Measurement Techniques and Standards of Electromagnetic Quantities from D-C to Millimeter-Wave Frequencies). Issued monthly. Annual subscription: $100.00 (Special rates for multi-subscriptions). Send subscription order and remittance to the Electromagnetic Metrology Information Center, Electromagnetics Division, National Bureau of Standards, Boulder, Colo. 80302